Lecture Notes in Geoinformation and Cartography

Series editors

William Cartwright, Melbourne, Australia
Georg Gartner, Wien, Austria
Liqiu Meng, München, Germany
Michael P. Peterson, Omaha, USA

About the Series

The Lecture Notes in Geoinformation and Cartography series provides a contemporary view of current research and development in Geoinformation and Cartography, including GIS and Geographic Information Science. Publications with associated electronic media examine areas of development and current technology. Editors from multiple continents, in association with national and international organizations and societies bring together the most comprehensive forum for Geoinformation and Cartography.

The scope of Lecture Notes in Geoinformation and Cartography spans the range of interdisciplinary topics in a variety of research and application fields. The type of material published traditionally includes:

- proceedings that are peer-reviewed and published in association with a conference;
- post-proceedings consisting of thoroughly revised final papers; and
- research monographs that may be based on individual research projects.

The Lecture Notes in Geoinformation and Cartography series also includes various other publications, including:

- tutorials or collections of lectures for advanced courses;
- contemporary surveys that offer an objective summary of a current topic of interest; and
- emerging areas of research directed at a broad community of practitioners.

More information about this series at http://www.springer.com/series/7418

Vasily Popovich · Christophe Claramunt
Manfred Schrenk · Kyrill Korolenko
Jérôme Gensel
Editors

Information Fusion and Geographic Information Systems (IF&GIS' 2015)

Deep Virtualization for Mobile GIS

Editors
Vasily Popovich
SPIIRAS 39
St. Petersburg
Russia

Christophe Claramunt
Naval Academy Research Institute
Brest naval
Brest
France

Manfred Schrenk
CORP—Competence Center of Urban
 and Regional Planning
Klosterneuburger
Austria

Kyrill Korolenko
NAVSEA
NUWC
Newport, RI
USA

Jérôme Gensel
Laboratory of Informatics of Grenoble
Université Grenoble Alpes
Saint Martin d'Heres
France

ISSN 1863-2246   ISSN 1863-2351 (electronic)
Lecture Notes in Geoinformation and Cartography
ISBN 978-3-319-16666-7   ISBN 978-3-319-16667-4 (eBook)
DOI 10.1007/978-3-319-16667-4

Library of Congress Control Number: 2015936362

Springer Cham Heidelberg New York Dordrecht London
© Springer International Publishing Switzerland 2015
This work is subject to copyright. All rights are reserved by the Publisher, whether the whole or part of the material is concerned, specifically the rights of translation, reprinting, reuse of illustrations, recitation, broadcasting, reproduction on microfilms or in any other physical way, and transmission or information storage and retrieval, electronic adaptation, computer software, or by similar or dissimilar methodology now known or hereafter developed.
The use of general descriptive names, registered names, trademarks, service marks, etc. in this publication does not imply, even in the absence of a specific statement, that such names are exempt from the relevant protective laws and regulations and therefore free for general use.
The publisher, the authors and the editors are safe to assume that the advice and information in this book are believed to be true and accurate at the date of publication. Neither the publisher nor the authors or the editors give a warranty, express or implied, with respect to the material contained herein or for any errors or omissions that may have been made.

Printed on acid-free paper

Springer International Publishing AG Switzerland is part of Springer Science+Business Media (www.springer.com)

Preface

This volume contains the selected papers and two invited papers presented at the *Seventh International Workshop on Information Fusion and Geographic Information Systems: Deep Virtualization for Mobile GIS (IF&GIS' 2015)* held during May 18–20, 2015 in Grenoble, France. IF&GIS is a biannual international workshop that brings together academics and industrials from a wide range of disciplines including computer science, geography, environmental and urban sciences, applied mathematics and social science at large. It continues a series of successful workshops held regularly since 2003 in St. Petersburg, in Brest in 2011, and 2013 back to the St. Petersburg. It has been continuously organized by "SPIIRAS Hi Tech Research and Development Office Ltd" and the French Naval Academy Research Institute. In 2015 IF&GIS was organized together with the University Grenoble Alpes in co-location with the W2GIS International Symposium. This year IF&GIS was specifically oriented towards virtualization issues related to mobile GIS.

Virtualization is nowadays an accepted standard IT practice and concept whose objective is to provide a set of software technologies that separate the device environment and the associated application software from the physical client device. When applied to the GIS domain, virtualization supports the development of self-contained, centrally managed and user-oriented GIS software on-demand. It allows users to be productive anywhere with virtual GIS-applications that work as if they are locally installed, extending the benefits of virtualization to satisfy a full range of the GIS user requirements.

The objective of this workshop was to present recent scientific and technological innovations related to the development of GIS that can be considered as part of this virtualization effort. The submission process attracted 23 full paper submissions from ten countries, amongst which ten were selected for full paper presentation and eight as abstract presentations at the workshop. The presented papers cover a large set of themes and have been scheduled in three sessions: Web-Based and Location-Based GIS Applications, Geospatial Crowdsourcing, Modeling and Computing, Applying GIS Technologies to Various Application Domains. The scope of the seventh IF&GIS workshop also covers multidisciplinary issues and applications such as environmental and disaster management, ethnography and historical studies. The program also features two invited papers by Dr. James D. Carswell from the Digital Media Centre of the Dublin Institute of Technology and Professor Leonid I. Borodkin from Lomonosov Moscow State University.

The success of the workshop was assured by the team efforts of sponsors, organizers, reviewers and participants. We would like to recognize the contribution of the Program Committee members and thank all reviewers for their support and hard work. Our sincere gratitude goes to all participants and all authors of the submitted papers. Finally, we also would like to acknowledge the support and collaboration of Springer's LNGC team, managed by Editorial Director Dieter Merkle, Publishing Editor Dr. Christian Witschel, Publishing Editorial Assistant Agata Oelschlaeger and Project Coordinator Suresh Rettagunta.

May 2015

Vasily Popovich
Christophe Claramunt
Manfred Schrenk
Kyrill Korolenko
Jérôme Gensel

Contents

Part I   Invited Papers

Design and Development of Personal GeoServices for Universities 3
Andrea Ballatore, Thoa Pham, Junjun Yin, Linh Truong-Hong
and James D. Carswell

**Spatial Analysis of Peasants' Migrations in Russia/USSR
in the First Quarter of the 20th Century**...................... 27
Leonid Borodkin

Part II   Web-Based and Location-Based GIS Applications

**RRM—A Referenced Routing Model to Generate a Semantic
Service of Navigation in Mobile Devices** 43
Ana María Magdalena Saldaña-Pérez, Miguel Torres-Ruiz,
Marco Moreno-Ibarra and Giovanni Guzmán-Lugo

Semantic Trajectories: A Survey from Modeling to Application...... 59
Basma H. Albanna, Ibrahim F. Moawad, Sherin M. Moussa
and Mahmoud A. Sakr

**Semantic Recommender System for Touristic Context
Based on Linked Data**.................................. 77
Luis Cabrera Rivera, Luis M. Vilches-Blázquez, Miguel Torres-Ruiz
and Marco Antonio Moreno Ibarra

Part III Geospatial Crowdsourcing, Modeling and Computing

Detecting Clustering Scales with the Incremental K-Function: Comparison Tests on Actual and Simulated Geospatial Datasets 93
Ran Tao, Jean-Claude Thill and Ikuho Yamada

Crowd Computing Framework for Geoinformation Tasks 109
Alexander Smirnov and Andrew Ponomarev

Dynamic Resources Management in Agile IGIS 125
Nataly Zhukova and Alexander Vodyaho

Part IV Applying GIS Technologies to Various Application Domains

Application of GIS Technologies in Historic and Ethnographic Research 149
Yan Ivakin and Vladislav Ivakin

Geoinformation Systems for Maritime Radar Visibility Zone Modelling 161
Oksana Smirnova and Vasily Svetlichny

Part I
Invited Papers

Design and Development of Personal GeoServices for Universities

Andrea Ballatore, Thoa Pham, Junjun Yin, Linh Truong-Hong and James D. Carswell

Abstract Personal GeoServices are emerging as an interaction paradigm linking users to information rich environments like a university campus or to Big Data sources like the *Internet of Things* by delivering spatially intelligent web-services. OpenStreetMap (OSM) constitutes a valuable source of spatial base-data that can be extracted, integrated, and utilised with such heterogeneous data sources for free. In this paper, we present a Personal GeoServices application built on OSM spatial data and university-specific business data for staff, faculty, and students. While generic products such as Google Maps and Google Earth enable basic forms of spatial exploration, the domain of a university campus presents specific business information needs, such as "What classes are scheduled in that room over there?" and "How can I get to Prof. Murray's office from here?" Within the framework of the *StratAG* project (www.StratAG.ie), an *eCampus Demonstrator* was developed for the National University of Ireland Maynooth (NUIM) to assist university users in

A. Ballatore
Center for Spatial Studies, University of California, Santa Barbara, USA
e-mail: aballatore@spatial.ucsb.edu

T. Pham
Department of Computer Science, National University of Ireland,
Maynooth (NUIM), Ireland
e-mail: thoa.phamthi@gmail.com

J. Yin
Department of Geography and Geographic Information Science,
University of Illinois at Urbana-Champaign (UIUC), Champaign, USA
e-mail: yinjunjun@gmail.com

L. Truong-Hong
Department of Civil Engineering, University College Dublin (UCD),
Dublin, Ireland
e-mail: linh.truonghong@gmail.com

J.D. Carswell (✉)
Digital Media Centre, Dublin Institute of Technology (DIT), Dublin, Ireland
e-mail: jcarswell@dit.ie

© Springer International Publishing Switzerland 2015
V. Popovich et al. (eds.), *Information Fusion and Geographic Information Systems (IF&GIS' 2015)*, Lecture Notes in Geoinformation and Cartography,
DOI 10.1007/978-3-319-16667-4_1

exploring and analysing their surroundings within a detailed data environment. This work describes this system in detail, discussing the usage of OSM vector data, and providing insights for developers of spatial information systems for personalised visual exploration of an area.

Keywords Personalised maps · Geoservices · Spatial-business data linking · OSM

1 Introduction

As today's spatially aware users are becoming more sophisticated and interested in retrieving more personalised information,[1] they will increasingly require more detailed two dimensional (2D) and three dimensional (3D) cityscapes linked to relevant non-spatial attribute data. In this work, *Personal GeoServices* are developed that employ mobile devices (spatially aware smartphones) to provide contextual search and visualisation utilities over 2D OpenStreetMap (OSM) building footprint data and detailed 3D building model data. At smaller map scales, Google Maps/Earth with satellite/street views can assist users searching for general information at specific locations. Users can search a street address on the map, then explore the location in street view mode, or find how to reach a certain address or location given options like pedestrian or driving constraints. However, generic query tools available in these familiar products are usually limited to keyword-based search. At larger local scales, where detailed 2D and 3D geometries and associated business data are needed, there is a recognised lack of advanced spatial search functionality and linked attribute information available in these products for domain-specific, task-oriented, and personalised visual exploration of an area [1].

For instance, the following types of questions (queries/searches) cannot be answered when interacting with Google Maps/Earth on a typical university campus: "What classes are scheduled in that room over there?"; "Whose office window is that up there?"; "What computer labs can I actually see around me from this location on campus?"; "What are their opening times?". In order to answer these types of task specific queries, Location Based Services (LBS) need the ability to search and link spatial map data together with non-spatial business data.

Spatial data includes detailed topology and geometry of objects while business data can describe the attributes or semantic aspect of related objects in some business domain. Conventional business data is often produced and managed by traditional enterprise information systems, often ignoring the spatial dimension altogether. However, business data can often be indirectly or virtually associated to spatial data via its location given by a generic address, room number, building name, or sometimes even geographic (lat/long) coordinates [1]. Linking spatial data

---

[1]http://googleblog.blogspot.ie/2013/01/mapping-creates-jobs-and-drives-global.html.

and business data together in one application can help to fulfil more task specific user needs In particular, decision-making applications need access to detailed local-scale data typically found in museums, hospitals, shopping malls, retail/office park settings, or a university campus. Business data specific to a university campus may be in the form of class schedules for a specific classroom, lists of equipment installed in a lab, office hours or contact details for a lecturer, today's special meal deal in the cafeteria, etc.

Typically, 2D vector "footprint" data provides just boundary geometry representation of physical objects (e.g. buildings, roads, rivers, etc.) in the horizontal plane. However, in our eCampus application, we link spatial and non-spatial attribute details in the vertical dimension as well for more advanced 3D information search operations. For instance, 3D model data of a building can include detailed digital representations of physical and functional characteristics for its different floors, rooms, windows and doors, where all objects are potentially available for interrogation.

In this paper, we describe our prototype eCampus information system in which the ideas above are implemented. Within the framework of the StratAG project (www.StratAG.ie), the *eCampus Demonstrator* was developed for the National University of Ireland Maynooth (NUIM), in collaboration with Dublin Institute of Technology (DIT), and University College Dublin (UCD). This Personal GeoServices application aims to assist users in exploring and analyzing their surroundings to answer more task specific user queries within a detailed data environment. The *Demonstrator* addresses two types of users: public users (e.g. visitors) and local users (e.g. students, faculty, and staff). Access privileges and query levels depend on user type. For example, visitors are presented with a campus map for general information querying about campus buildings and rooms. Visitors are also provided with general campus news, events, and utilities for navigation to various buildings or rooms, or any other locations on campus using different routing options. In addition to these general functions, staff and students are able to overlay on the map their individual class schedules together with personalised news feeds and events tailored to their academic and social interests.

Each project partner in the StratAG cluster is responsible for developing different functionality in the *eCampus Demonstrator*, such as; utilities for 2D/3D directional and visibility-based querying (DIT), path navigation assistance (NUIM), personalised news and events according to user interests (UCD), or for detailed mapping and modelling of the campus infrastructure itself (NUIM/DIT). RESTful webservices [2–4] were chosen as the deployment technology for these distributed components due to its simplicity when applied to the geospatial domain [5]. Regarding 3D maps, there are several commercial and free mapping products available that allow users to incorporate 3D building models for such visualization and interaction type applications. Of these, Google Earth (GE) was selected for displaying the 3D building models in this work because it is both free and increasingly familiar to web and wireless GeoService users, although currently GE does have voluminous data processing limitations that must be addressed to accommodate real-time display.

In the remainder of the paper, we first discuss some related work before introducing our approach to integrating 2D OSM and 3D model data within the application. The 2D/3D building information is then imported into a spatial database and converted to GE readable format in the case of 3D map display. Then we present the system architecture of the *eCampus Demonstrator* based on a Resource Oriented Architecture (ROA) model. Some unique search functionalities of eCampus are described in detail together with its graphical user interfaces. Finally we draw conclusions and give some possible direction for future work.

2 Related eCampus Applications

Assisting people with exploring an area, such as a university campus, with a mapping interface is very useful, and is not a new idea in itself. Some existing projects provide this type of GeoService, as listed below.

Kent State University Campus Maps [6] is a web-based application providing an interactive 2D map with detailed information and images of each building in the campus. It also highlights specific locations on the campus such as computer labs, parking, sculpture walk, and residence halls. The *Get Direction* functionality allows finding pedestrian/driving/cycling routes between two locations, e.g. using building names. However, this application provides very basic query/search functionality overall, has no 3D maps, or any form of personalisation.

The Interactive Map developed in University of California, Berkeley [7] provides an isometric campus map with clickable buildings to provide detailed information and an image of each building. There is also a list of building names placed outside the map window from where users can choose a name, then the corresponding building object on the map is highlighted. However, this application has no search facility, no routing service, and provides limited spatial support.

The University College Dublin Mobile services [8], released natively on Android and iOS, has the following functionalities: Campus maps, details about places, and tours; Campus directory; Access to Library; Access to e-learning facility (Blackboard); Schedule of general events (lectures, concerts, etc.); Campus news; Image search on university archives; and Emergency numbers. The application has two main versions, one for staff and one for students, with different permissions. Although this application provides many useful services, particularly in relation to campus events, its spatial support is still very limited. It has only a simple, non-interactive 2D map, with no dynamic routing functionality.

In [9], the early stages of a web-based campus information system was developed for the University of Karabuk, Turkey, allowing users to explore the university campus in 3D. It provides information at the building level and points of interests, but room level details have not been fully incorporated, and the implementation does not provide utilities to further query the area beyond its physical, spatial nature.

The Youngstown State University developed interactive 2D and 3D campus maps [10] which allows the retrieval of building information when clicking on a building or to download a KMZ file (zipped KML archive file) of an area to interact with in Google Earth. However, the attribute information provided is also limited to building level only.

A clear trend is evident in the above GeoService applications in that more advanced functionality for exploring non-spatial business data of sub-objects like rooms, external windows and doors to retrieve the content, schedule, or purpose of a specific room in a building is still obviously lacking. This is largely due to insufficient levels of geometric granularity of building models available in today's online mapping platforms like Google Earth, and importantly a subsequent lack of any spatially linked business data.

3 OSM and Volunteered Geographical Information (VGI)

The availability of detailed geographic data is critical for delivering comprehensive GeoService applications. Most geospatial data in Europe is collected and controlled by either national mapping agencies (e.g. OSi in Ireland) or private companies, such as Google Maps, Yahoo! Maps, and Bing Maps. However, data coverage over many areas can still be of considerably poor quality and extent—especially in less populated areas.

Some initial research to address this shortcoming looked at imaging and geo-referencing public displays of "You Are Here" type maps to fill the coverage gap for local navigation purposes [11]. However, a rapid growth in volunteered geographic information (VGI) has also started to fill this gap with OSM being a successful example of this. In relation to VGI, this "citizen-as-sensor" paradigm contributes to OSM by creating, assembling and disseminating geospatial features including streets, highways, buildings, etc. and gradually this collective geospatial information shows surprising coverage all over the world [12, 13]. Another vitally important feature of using OSM data is that *"you are free to copy, distribute, transmit and adapt our maps and data, as long as you credit OpenStreetMap and its contributors"*,[2] which affords developers to build value-added applications on top of it.

Considering this increasing trend of free VGI sourced data, our project is built upon base geospatial data from OSM covering the National University of Ireland, Maynooth (NUIM) and its surrounding areas. A screenshot of the OSM map coverage for this area at the time of project implementation is shown in Fig. 1a, while the corresponding map coverage over the same time/area from Google Maps is shown in Fig. 1b. The considerably more detailed OSM data is mainly created by students from NUIM and is freely available for inclusion in all value-added projects, a clear example of how OSM grows research and business opportunities through volunteers contributing data [14].

In this project, 2D footprints of NUIM campus buildings were downloaded directly from the OSM map interface (www.openstreetmap.org) and/or from its data

---

[2]http://www.openstreetmap.org.

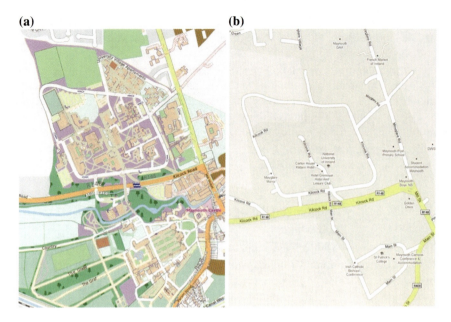

Fig. 1 **a** 2012 OSM map coverage of NUIM, Ireland **b** Map coverage of the same place and time from Google Maps

repository (e.g. *Planet OSM*[3] and *Metro Extracts*[4]). For NUIM campus data, which covers a relatively small area, the entire campus can be exported by drawing a bounding box over the campus region to specify the area of interest. The downloaded OSM data contains several layers of geometry types (features), such as polygons (e.g. building footprints), polylines (e.g. roads), and points (e.g. points-of-interest). In our case, a layer of building polygons was extracted and inserted into an Oracle Spatial database via the Feature Manipulation Engine (FME) Workbench utility [15], which is a toolset that converts and transfers data between different data formats. The inserted polygon data in Oracle Spatial is shown in Fig. 2.

4 Preparing 3D Models for Google Earth Integration

Google Earth (GE) allows developers to upload 3D building models for direct visualization in the GE environment or for inclusion in a webpage using their application programming interface (API). When retrieving data for visualization, a client (desktop or mobile) queries for data either from temporary cache memory or

---

[3]http://planet.openstreetmap.org/.
[4]http://metro.teczno.com/.

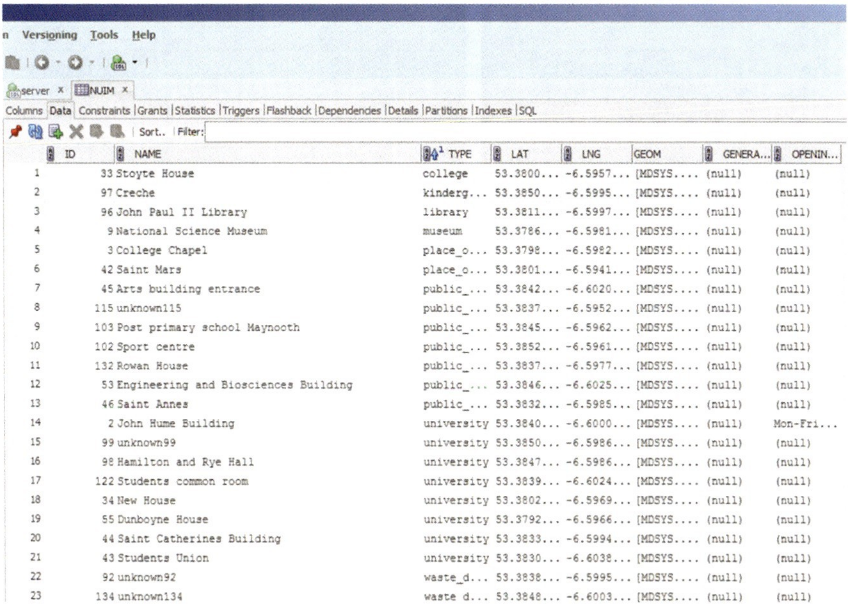

	ID	NAME	TYPE	LAT	LNG	GEOM	GENERA...	OPENIN...
1	33	Stoyte House	college	53.3800...	-6.5957...	[MDSYS....	(null)	(null)
2	97	Creche	kinderg...	53.3850...	-6.5995...	[MDSYS....	(null)	(null)
3	96	John Paul II Library	library	53.3811...	-6.5997...	[MDSYS....	(null)	(null)
4	9	National Science Museum	museum	53.3786...	-6.5981...	[MDSYS....	(null)	(null)
5	3	College Chapel	place_o...	53.3798...	-6.5982...	[MDSYS....	(null)	(null)
6	42	Saint Mars	place_o...	53.3801...	-6.5941...	[MDSYS....	(null)	(null)
7	45	Arts building entrance	public_...	53.3842...	-6.6020...	[MDSYS....	(null)	(null)
8	115	unknown115	public_...	53.3837...	-6.5952...	[MDSYS....	(null)	(null)
9	103	Post primary school Maynooth	public_...	53.3845...	-6.5962...	[MDSYS....	(null)	(null)
10	102	Sport centre	public_...	53.3852...	-6.5961...	[MDSYS....	(null)	(null)
11	132	Rowan House	public_...	53.3837...	-6.5977...	[MDSYS....	(null)	(null)
12	53	Engineering and Biosciences Building	public_...	53.3846...	-6.6025...	[MDSYS....	(null)	(null)
13	46	Saint Annes	public_...	53.3832...	-6.5985...	[MDSYS....	(null)	(null)
14	2	John Hume Building	university	53.3840...	-6.6000...	[MDSYS....	(null)	Mon-Fri...
15	99	unknown99	university	53.3850...	-6.5986...	[MDSYS....	(null)	(null)
16	98	Hamilton and Rye Hall	university	53.3847...	-6.5986...	[MDSYS....	(null)	(null)
17	122	Students common room	university	53.3839...	-6.6024...	[MDSYS....	(null)	(null)
18	34	New House	university	53.3802...	-6.5969...	[MDSYS....	(null)	(null)
19	55	Dunboyne House	university	53.3792...	-6.5966...	[MDSYS....	(null)	(null)
20	44	Saint Catherines Building	university	53.3833...	-6.5994...	[MDSYS....	(null)	(null)
21	43	Students Union	university	53.3830...	-6.6038...	[MDSYS....	(null)	(null)
22	92	unknown92	waste_d...	53.3838...	-6.5995...	[MDSYS....	(null)	(null)
23	134	unknown134	waste d...	53.3848...	-6.6003...	[MDSYS....	(null)	(null)

Fig. 2 2D footprints (polygons) of NUIM buildings inserted into Oracle Spatial

from GE databases. However, due to real- time display requirements and other issues, GE can only process very simple 3D block models if display speed is a priority. Therefore, in order to include our detailed 3D campus building models in GE, we first needed to develop a modelling workflow to transform raw LiDAR point cloud data to Google Earth KML format using various mapping tools like CloudWorx [16] and the FME Workbench utility. The KML format supports both solid and polygon data types, and both are needed to address real-time GE display issues as discussed below.

Raw data used to construct detailed building models for 3D city maps can be obtained from various sources through a wide range of techniques. With recent developments in photogrammetry and remote sensing, building models can be automatically reconstructed given the geometric resolution of satellite imagery and Light Detection and Ranging (LiDAR) point cloud data [17]. The geometric and semantic properties of 3D models are typically stored in five consecutive levels-of-detail (LoD), in which LoD0 defines a coarse regional scale model while LoD4 denotes architectural building models with detailed walls, roof structures, balconies, interior structures, and detailed vegetation and transportation objects [18]. An advantage of the LoD approach is the coherent modelling of semantics and geo-metric/topological properties together at each level, where geometric objects get assigned to semantic objects. In order to meet our eCampus objectives, 3D building models have to be constructed at least to LoD3 level.

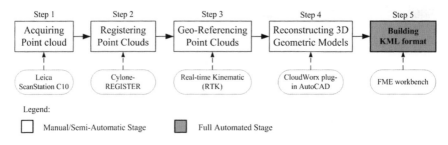

Fig. 3 LiDAR workflow for 3D BIM modelling

To achieve this LoD, a modelling workflow (Fig. 3) was developed to reconstruct building models suitable for real-time GE display from LiDAR point clouds. LiDAR point clouds of campus buildings were first acquired using a Leica ScanStation C10 controlled by Cyclone-3D Point Cloud Processing Software installed on a laptop linked to the scanner. Subsequently, the point clouds were registered and geo-referenced within the Cyclone environment [19] and the 3D building models manually created using AutoDesk and a CloudWorx plug-in [16]. Leica CloudWorx for AutoCad offers many manipulation and editing tools to assist users to trace or auto fit lines, arcs and polylines to 3D point cloud data. Finally, by employing FME Workbench, a 3D building model's underlying CAD geometry is transformed to KML format to allow for online display by web-based mapping applications in Google Earth. A more detailed description of each step in this modelling workflow can be found in [20].

This modelling methodology was applied to the north campus of the NUIM study area. This area was selected because it contains a mixture of simple and complex buildings with various architectural styles (e.g. historic and modern buildings), which can be most problematic when reconstructing 3D building models. The tallest building is 20 m. Due to GE constraints with how it processes (accesses/displays) 3D solids, the AutoCAD DWG format solid models created in Step 4 (above) have some limitations after converting to 3D solid KML format concerning access to their object/sub-object semantic attributes (such as building/room name).

In Google Earth, linked attribute information cannot be accessed by simply clicking (pointing/selecting) just anywhere on the solid building shape. To overcome this limitation, the solid building shape must be converted to 3D polygon data, which can be accessed in GE by clicking anywhere inside the polygon. However, by transforming all 3D solids to KML polygons, a map scene can contain a huge number of polygons because numerous polygons are required to represent all the external doors/windows, which can make the KML file so large that it exceeds the limit that GE can support for real-time display. Therefore, the KML 3D polygon file has to be thinned to show important detail of attribute-bearing objects only. So, an overlay layer of empty 3D polygons around each window/door was added and the DWG solids of the doors/windows were transformed to KML solids for visualisation purposes, otherwise users would only see (un-selectable) holes in a wall rather than (clickable) doors or windows.

In summary, 3D solids and 3D polygons together represent the campus building models in KML format with their associated rooms linked to any available metadata information. While polygons are applied to all objects, 3D solids are also used for visualising geometries of window/door frame details. This allows us to assign a different appearance to each building sub-component by filling with either a colour or texture and importantly the ability to click anywhere inside a polygon shape (e.g. window/door/wall) to select/query the object directly.

5 *eCampus* Architecture

The *eCampus Demonstrator* is a browser-based application based on 2D OSM data and 3D GE data and is accessible to both desktop and mobile devices. It aims to help users explore in more detail the campus by providing them maps and utilities for both 2D and 3D querying and visualisation. Different search functionality is provided so that users can ask questions by interacting with the map itself. For instance, they can ask: "What is that building over there?" by pointing at it with their mobile device; "What is the class schedule of this room?" by clicking on its window in the mobile/desktop display or by choosing a room ID from a list; "What can I actually see around me?" when standing at a particular location on campus; they can also ask to visualise a route together with directional images (i.e. containing superimposed arrows pointing the way) between two buildings/rooms or any location by choosing a building/room from a list or by clicking locations directly on the map. Query results are visualised on the 2D/3D map overlaid with further university specific business data where available. The application architecture includes 3 layers: interface layer, web-services, and database layer (Fig. 4).

5.1 Database Layer

The 2D building footprint and detailed 3D models of NUIM campus that serve the spatial data web-services are physically hosted at DIT in Oracle Spatial 11g databases. Other data related to pathways, roads and building images are hosted at NUIM in PostGIS databases. Many 2D campus footprints were first downloaded from OSM and then uploaded to the Oracle Spatial DBMS where geometry data is stored in a single column data type of *SDO_GEOMETRY* to define the geometry type (e.g. points, lines, polygons, solids, etc.), the dimension, and an array of x,y (and z for 3D) coordinates comprising points or vertices of campus objects.

In Sect. 4, the workflow for 3D BIM modelling and export to KML format for displaying in GE was discussed. In fact, 3D building detail is also needed and stored in Oracle Spatial for visibility calculations and other advanced search operations. In this case, the FME workbench utility was also used to export the 3D models to the spatial database—a similar process as when exporting to KML. After

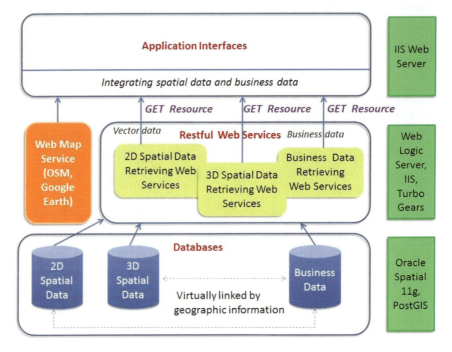

Fig. 4 Three Tier *eCampus Demonstrator* System Architecture

reaching Step 4 (3D BIM in AutoCAD) in the modelling workflow, there are two groups of layers stored in the AutoCAD database: (i) stored solid components involving exterior walls, window/door frames, roofs, and balconies and (ii) stored polygons of window/door extents. These two layers are then exported to Oracle Spatial. As mentioned previously, the reason for this type of data management is because of "clicking" restrictions inherent to Google Earth regarding pointing/selecting solid building objects/sub-object attributes directly from the map.

5.2 Web-Services Layer

At the logical level, the spatial data retrieval web-services and business data retrieval web-services are installed in this layer.

5.2.1 Spatial Data Retrieval Web-Services

Spatial data retrieval web-services include routing navigation, image retrieval for directional visualisation of routes, and 2D and 3D visibility based directional

Design and Development of Personal GeoServices for Universities 13

Fig. 5 The directional image web-service result overlaid on the OSM base map. Clicking on each thumbnail opens a larger image showing an *arrow* pointing the direction to travel

querying; 2D Isovist, point-to-select, and field-of-view queries; plus 3D Isovist, point-to-select, and frustum spatial searches. These web-services were developed by various project partners on different platforms, but have the same deployment methods in the form of *RESTful* web-services [2, 3]. More specifically, an IIS (Internet Information Services) server is appointed to host the web-services, and query requests are constructed using standard HTTP calls containing a valid URL filled with the required query parameters.

The routing navigation web-service needs query parameters such as transport mode (e.g. pedestrian, driving, wheelchair, or directional images), and a starting and destination location in terms of longitude and latitude coordinates taken from clicking on the map. The navigation web-service then returns a list of points (i.e. OSM object vertices along the route) in KML format. The *eCampus* applications then uses OpenLayers API to read this list and connect the KML points to display the route as a line drawn on the map. Users can also provide a building or room name for the start/end location. The corresponding coordinates of the building/room are retrieved from 2D and 3D spatial databases and passed to the web-service in this case.

The directional images web-service provides a list of images (thumbnails) along the route augmented with superimposed arrows pointing the way. The parameters for this web-service are the same as for normal routing. The web-service returns in JSON format a list of image URLs, their location, and the view angle of each image with respect to the direction (path) needed to follow. Figure 5 shows how the directional image web service result displays on the OSM basemap.

2D and 3D visibility-based directional query web-services correspond to different search options in relation to different spatial data types in the database. These web-services are divided into two sub-groups, one applied to 2D OSM building footprints and the other applied to the 3D GE (KML) building models. 2D Isovist view, 2D Field-of-View, 2D/3D Line-of-Sight (Point-to-Select), 3D Frustum and

Fig. 6 The 2D Isovist web service result overlaid on the OSM basemap. Only objects that a user can actually see out to a specified distance (e.g. 100 m) in 2D get returned by the query

3D Threat Dome are the different types of 2D and 3D spatial queries available. More detail on these spatial search algorithms and the web-services developed for each can be found in [21, 22] (Fig. 6).

5.2.2 The Business Data Retrieval Web-Services

Useful university business data attributes for student/staff users mainly relates to classroom schedules and facilities. *RESTful* web-services were developed to retrieve this information by querying an external NUIM database containing university business data associated to room numbers. The web service returns the classroom schedule in html format.

5.3 Interface Layer

The interface layer consists of html pages displayed to users (client side) in a standard web browser. At this level, spatial data returned from 2D queries is visualised on 2D OSM maps as additional layers using OpenLayers API. For 3D queries, the results are returned in JSON format and drawn as placemarks added to a GE view. The integration of spatial data and business data is performed at the client side when attribute information about buildings and associated rooms/objects is found.

Design and Development of Personal GeoServices for Universities

6 Advanced GeoService Functionality

The eCampus Demonstrator is developed for both web and wireless devices. However, there are some differences in the look and feel of the interfaces and user interactions between the desktop version and the smartphone version. This is mainly due to additional limitations of the Google Earth API for mobile devices. In addition, on mobile devices information like user location, tilt, and compass readings can be captured automatically, while users need to input them when interacting with the desktop interface. The *eCampus Demonstrator* GeoService functionality and its organisation is depicted in Fig. 7.

6.1 Exploring eCampus Information

This search functionality is available from the home page of the desktop application (Fig. 8). It shows a 2D OSM basemap of the campus where users can selectively click on each building (polygon) on the map to explore more detailed information visualised on the right side of the interface. The available business data includes information on rooms within the building, faculties/departments in each building, opening hours of the building, car park information, building images and detailed architectural plans (Figs. 8 and 11).

6.1.1 Personalised Information

This functionality provides information specifically targeted to user interests. Campus users manage their interests by adding or removing them from their profile.

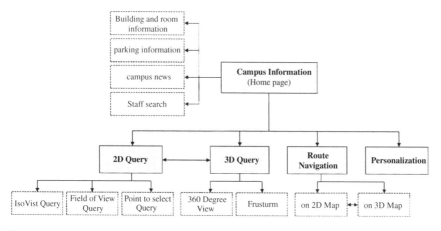

Fig. 7 *eCampus* search functionality. There are 5 main search functions in the *eCampus Demonstrator*. Each function provides a further subset of query utilities (*dashed boxes*)

Fig. 8 Selected building information displayed on right side of desktop interface

Fig. 9 Personalised news and events overlaid on OSM basemap

An auto complete list of keywords is provided to help users in selecting and adding interest keywords. After logging into the system with a student/staff ID, campus information such as school/course calendar and news/events are personalized based on these chosen interests (e.g. History, Sports, Societies, Restaurants, etc.) (Fig. 9).

6.2 2D Queries on the Desktop

2D search functionality allows us to perform Isovist, Field-of-View and Point-to-Select queries. On the desktop interface, a user first chooses a query type and then clicks a location and drags the mouse to indicate the search radius directly on the OSM map. Depending on the query type, the user may optionally enter a field-of-view angle and search direction. The query result is displayed as blue shaded

Fig. 10 Results of 2D Field-of-View query looking northwards out to 160 m

buildings with the field-of-view (viewshed) shaded in red (Fig. 5). As opposed to a simple range query where all objects out to a certain distance are retrieved, a 2D Isovist is a 360° line-of-sight query around a user's location that is clipped by OSM building footprints. This allows for more task specific queries such as "retrieve all objects around me that I can actually see within 50 meters".

By further constricting the 360° Isovist search, the Field-of-View query allows users to indicate a viewshed angle, direction, and distance from their chosen location. The application then retrieves all visible objects that fall within that angular Field-of-View, in that particular direction, and within that specified distance (Fig. 10).

The Point-to-Select query intends to mimic a user pointing their smartphone at actual objects in the real-world, but is invoked on the desktop by clicking 2 (or more) locations on the map and retrieving all objects that intersect the line (Fig. 11).

Fig. 11 2D Point-to-Select query with 4 points indicating the chosen "line-of-sight"

6.3 2D Queries on the Mobile

Mobile interfaces for exploring the campus on a smartphone are depicted in Fig. 12. Building information gets retrieved by simply tapping on any building in the OSM map.

The mobile interface takes into account the current user location from the device GPS, tilt of the device from accelerometer, and azimuth of the pointing direction from the digital compass. The search radius is indicated by dragging the distance slider left or right with the resulting red viewshed changing dynamically to indicate scale. The final viewshed query shape gets clipped by the OSM building footprints before being sent to the spatial database to retrieve any intersected objects (Fig. 13).

Regarding the mobile version of route navigation, users can ask to find a route from their current location (based on their GPS location) to any building/room or between any two campus buildings selected from a list (Fig. 14).

6.4 3D Queries on Desktop and Mobile

The 3D eCampus search functionality displays NUIM maps in KML format using Google Earth (3D map). At this stage, users can query the campus by clicking directly on any building or external door/window (i.e. room). The corresponding building/room information gets overlaid on the 3D map (Fig. 8). The *eCampus Demonstrator* provides two 3D search options: 360° Isovist query (Threat Dome) and a directional Frustum query. The 360° Isovist query is a 3D version of the 2D Isovist query, which means it returns objects that users can actually see around them in all directions horizontal and vertical out to a specified distance (Fig. 15). The

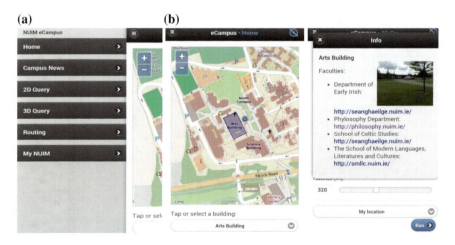

Fig. 12 Mobile eCampus interface on smartphone. **a** Main menu. **b** Building query

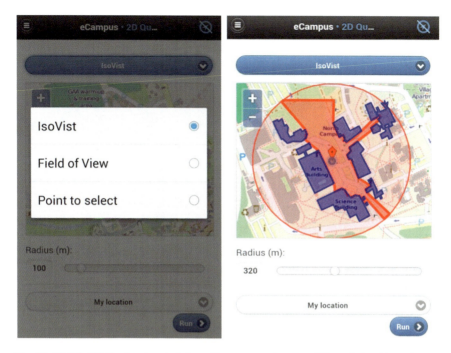

Fig. 13 Mobile 2D Isovist viewshed (in *red*) set to search out to 320 m. Only objects that can actually be seen (in *blue*) from current location get returned

final dome shape (i.e. search space) is determined by first calculating all intersections between a sphere centred on the user's location and the 3D building models stored in the spatial database. Next, the intersection points are joined into a 3D polygon shape that is then used as the query "window" into the database to retrieve all visible objects (e.g. rooms/buildings) it contains.

The Frustum view query is the 3D version of 2D Field-of-View query. It constrains the shape of the search space to a location having vertical and horizontal angular visibility (and tilt) in a particular direction, out to a specified distance and clipped to the 3D building models stored in the spatial database. The intersection of the resulting frustum query "window" with other database objects represents what users can actually see in a given 3D direction, constrained, for example, to their actual field-of-view (Fig. 16).

Currently, due to API limitations with mobile GE, e.g. in drawing/inserting user defined 3D objects, the mobile 360° Isovist and Frustum query interfaces look the same as in 2D. However, the query results do take into account the true 3D nature of the search space. In this case, the search results provide a list of buildings and rooms intersected by the 3D query but displayed on a 2D OSM map view (Fig. 17) (i.e., the objects intersected by the yellow Threat Dome shape in Fig. 15 or by the green frustum shape in Fig. 16).

Fig. 14 Mobile route navigation with directional images between selected buildings/rooms. **a** Pedestrian walk. **b** Routing with directional images

Fig. 15 3D 360° Isovist (Threat Dome) view query. Only visible objects (e.g. buildings/rooms) that fall within the dome shape get returned by the query

Design and Development of Personal GeoServices for Universities   21

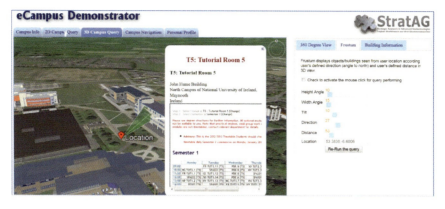

Fig. 16 3D Frustum query. The final frustum shape is clipped by any building objects in its path before getting utilised as a 3D query window in the spatial database

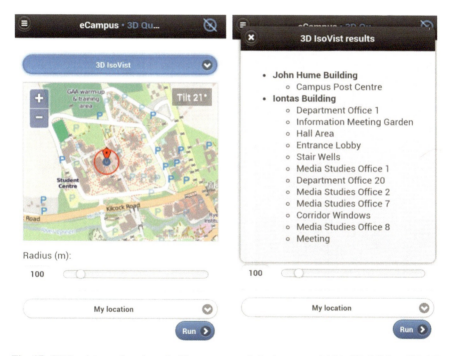

Fig. 17 3D Isovist search and result. Note query result is shown overlaid in 2D OSM as GE API does not yet allow functionality for user-defined customised 3D objects overlaid on their mobile 3D maps

7 Discussion

During the development phase of this application, some important limiting factors emerged regarding the technology employed. After first analyzing the capabilities of the chosen technologies for implementation, the following discussion presents some advantages and disadvantages of our approach.

7.1 3D Modelling and Visualisation

When creating 3D virtual cities for Personal GeoService applications and for general mobile spatial interaction in the *Internet of Things*, the most important task is to generate detailed and geometrically accurate building models. In contrast to traditional methods (e.g. on-site surveying), terrestrial laser scanning is an attractive alternative for collecting building coordinate data in terms of field time and accuracy [23–25]. However, the process of building detailed 3D models from point cloud data is still quite a manual process. In our application, it takes around 6–8 h for each building. This implies automatic or semi- automatic processes must be developed to reconstruct building models with LoD3 to reduce post-processing bottlenecks for city-wide 3D modelling workflows used in this type of application.

As the GE platform currently only displays 3D objects above the earth's surface, this restricts visualisation of any underground detail below ground elevation level. To display and interact with entire building models above and below ground, an alternate web mapping platform should be considered. Also, mobile device browsers do not currently (2013) support the Google Earth API, therefore there are limitations when visualising 3D models and other customised vector objects on smartphones.

7.2 Integration Level of Spatial Data and Business Data

As mentioned previously, the integration of spatial data and business data is performed at the client side, i.e. at the visualisation level. We considered three options to provide spatial data and related business data to users as shown in Fig. 18.

(a) Early-integration: In this approach, retrieving business data is performed from inside the spatial data retrieving WS. The final results sent back to GUI include spatial data and business data.
(b) Aggregated web-service: A new web-service is developed to compose the results returned by the spatial data web-service and the business data web-service.
(c) Integration at the visualisation level: The results of the spatial data web-service and business data web-service are overlaid at the visualisation level. We have chosen the third approach: integration at the visualisation level, as this approach provides some advantages compared to the other two approaches (Table 1).

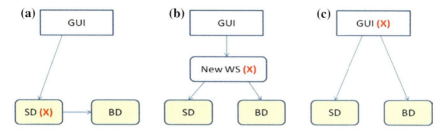

Fig. 18 Different spatial data and business data integration approaches. The *red* (X) represents the integration point. The *arrow* describes the calling direction. *WS* Web-Service; *SD* spatial data WS; *BD* business data WS; *GUI* Graphical User Interface

Table 1 Spatial data and business data integration options

(a) Early-integration	There is a dependency of SD retrieving WS and a specific BD retrieving WS. That means SD WS cannot be reused for other purposes
(b) Aggregated web-service	SD and BD WS are independent. It depends on the user needs that the aggregated WS (AggWS) will integrate suitably SD and BD WS. Suppose $fA(WS)$ is the cost of analyzing the result of a WS, $fC(WS)$ is the cost of calling a web-service, then the cost of this approach to display the result is: $fC(AggWS) + fA(Agg\ WS) + fC(SD) + fA(SD) + fC(BD)$
	Note that the results of AggWS needs to be analyzed to draw the geometry shapes and the business data is then added to the feature data of the geometry object. In case the results of AggWS is in KML format, there is no need of analysing AggWS (so no cost), all spatial data and business data can be visualized. However, in that case the visualisation is fixed according to the API provided
(c) Integration at visualisation level	SD and BD WS are independent. It depends on user needs that suitable SD and BD WS are consumed at the visualisation (client side code source). The cost to display the result to the users is: $fC(SD) + fA(SD) + fC(BD)$
	Visualisation of the results is flexible according to the users' needs

7.3 OGC Services

The Open Geospatial Consortium (OGC) has developed some standards for geospatial processing technologies to enable applications from different commercial vendors to interoperate. However, the locationing services developed within OGC focus mainly on tracking and location-based applications for mobile devices [26]. These services are far from what we require in this application, where all *RESTful* web-services for location-dependent directional and visibility space querying have been developed in our own algorithms.

7.4 ROA Instead of SOA

For the last decade, Service Oriented Architectures (SOA) have been widely used for distributed applications, particularly on the Web. In Geographic Information Systems (GIS), there is no exception here. For instance in [27], a GIS web-service architecture was proposed based on SOA technology. However, while SOA is a proven approach, in some cases it can be overly complicated and processor heavy. For example, when handling a SOAP message, the client (desktop or mobile) needs to send a request with parameters constructed and wrapped in XML format with special headers and other elements. It also has to parse any response from the server in the same effusive XML format [28].

In the case where the client is a mobile device, this approach contains far too much processing overhead in terms of the volume of data, most of it quite unnecessary, that must be sent/received on mobile devices having relatively limited wireless connection speeds and often a data transmission cost [29]. In this respect, JSON is a much lighter data format in terms of processing and transmitting wirelessly. Furthermore, we agree with the general statement that the *REST* architecture provides a *"scalable and simple deployment of web-services and particularly appealing for Earth and Space Science"* [30] as *RESTful* web-services have been much used in geo-information sharing.

7.5 Dependency of 3D Query Performance and 3D Data Details

In our application, users carry out spatial queries from outside of buildings. Therefore only the geometries of exterior structural components of the building (e.g. facades, roofs, windows, doors, balcony, canopy, etc.) associated with room level attribution (e.g. room name and function) were loaded into the database. In this way, it helps to reduce the complexity of the 3D models and thus improve 3D spatial query performance.

8 Conclusions

Providing users with business context data in location dependent queries helps to fulfil more task specific user needs within detailed data environments. In order to meet that objective, there is a need for 3D building modelling to achieve at least LoD3. The workflow employed in this paper was successful in reconstructing geometrically accurate building models with LoD3 detail. However, the procedure is time consuming for larger project areas where numerous building models need to be reconstructed; therefore, automation of this approach is still an open problem. The flexibility, interoperability and heterogeneity of this kind of GeoService application demands a suitable software architecture. In particular to this geospatial application, a Resource Oriented Architecture (ROA) was chosen for the implementation.

At time of writing, the *eCampus Demonstrator* presented in this paper is among the first GeoService applications to explore an area in detail on both desktop and mobile 2D and 3D maps. It provides users with more personalised search utilities, like directional/visibility query functionality, than contemporary eCampus type information systems currently allow. However, there are still some significant 3D data processing limitations that need solving by GE if their mapping platform is to be widely adopted for similar detailed data geo-application development in future. In the meantime, alternative platforms to consider could be 3D games engines such as *Unity*.[5] We also need more testing with student/staff users to get their feedback on overall functionality as well as the general performance of our application. Semantic Web technologies might also be employed to facilitate the integration of heterogeneous data [31]. In summary, this detailed data eCampus implementation can be considered as a starting point for developers and researchers when developing for similar application domains, such as business parks, hospitals, airports, and shopping districts.

Acknowledgments Research presented in this paper was funded by a Strategic Research Cluster Grant (07/SRC/I1168) by Science Foundation Ireland under the National Development Plan. The authors gratefully acknowledge this support.

References

1. Pham Thi TT, Truong-Hong L, Yin J, Carswell J (2013) Exploring spatial business data: A ROA based eCampus application, W2GIS 2013. LNCS, vol 7820. Springer, Berlin, pp 164–179
2. Guinard D, Trifa V, Wilde E (2010) A resource oriented architecture for the web of things. In: IEEE international conference on internet of things, Tokyo
3. Lucchi R, Millot M, Elfers C (2008) Resource oriented architecture and REST. JRC scientific and technical report
4. Fielding RT (2000) Architectural Styles and the Design of Network-based Software Architectures. PhD dissertation, University of California, Irvine
5. Kurtagic H, Birch J, Zeiss G (2009) An open architecture for RESTful geospatial web-services. FOSS4G Sydney
6. Kent State University campus maps. http://www.kent.edu/campuses/maps/-map.cfm. Accessed May 2014
7. University of California, Berkeley Interactive Map. http://www.berkeley.edu/map/3dmap/3dmap.shtml. Accessed May 2014
8. University College Dublin Mobile services. http://www.ucd.ie/itservices/-%20itsupport/mobileservices. Accessed May 2014
9. Kaharaman I, Karas IR, Rahman AA (2011) Developing web-based 3D campus information system. ISG and ISPRS

---

[5]http://unity3d.com/unity.

10. YSU 3D campus map. http://www.ysu.edu/campusmap/. Accessed May 2014
11. Schöning J, Krüger A, Cheverst K, Rohs M, Löchtefeld M, Taher F (2009) PhotoMap: using spontaneously taken images of public maps for pedestrian navigation tasks on mobile devices. ACM, Bonn, Germany
12. Goodchild MF (2007) Citizens as sensors: the world of volunteered geography. GeoJournal 69:221
13. Haklay M (2008) And Weber, P. OpenStreetMap—User generated street map. IEEE Pervasive Comput 2008:12–18
14. Jacob R, Zheng J, Ciepłuch B, Mooney P, Winstanley AC (2009) Campus guidance system for international conferences based on OpenStreetMap. In: Proceeding W2GIS '09, Maynooth, Ireland. Springer, pp 187–198
15. Safe software (2013) FME desktop
16. Leica Geosystems AG (2011) Leica CloudWorx for AutoCAD
17. Haala N, Kada M (2010) An update on automatic 3D building reconstruction. ISPRS J Photogr Remote Sens 65:570–580
18. Gröger G, Kolbe TH, Nagel C, Hafele KH (2014) OpenGIS City Geography Markup Language (CityGML) Encoding Standard (OGC 12-019). Version 2.0.0. OGC 12-019. Open Geospatial Consortium. http://www.opengeospatial.org/standards/citygml. Accessed May 2014
19. Leica Geosystems AG (2011) Leica Cyclone-3D point cloud processing software
20. Truong-Hong L, Pham Thi TT, Yin J, Carswell J (2013) Detailed 3D building models for Google Earth integration. ICCSA 2013. Springer, Ho Chi Minh city
21. Carswell JD (2010) 3DQ: Threat dome visibility querying on mobile devices. GIM Int 24(8):24
22. Carswell J, Gardiner K, Yin J (2010) Mobile visibility querying for LBS. Trans GIS 14(6):791–809
23. Truong-Hong L (2011) Automatic generation of solid models of building Façades from LiDAR data for computational modelling. School of Architecture, Landscape and Civil Engineering. PhD University College, Dublin
24. Truong-Hong L, Laefer DF, Hinks T, Carr H (2012) Flying Voxel method with Delaunay triangulation criterion FOR Façade/feature detection for computation. ASCE J Comput Civil Eng 26:691–707
25. Truong-Hong L, Laefer DF, Hinks T, Carr H (2012) Combining an angle criterion with voxelization and the flying Voxel method in reconstructing building models from LiDAR Data. Computer-Aided Civil Infrastruct Eng. doi:10.1111/j.1467-8667.2012.00761.x
26. OGC Report (2007) Summary of OGC Web-services, Phase 4, Interoperability Testbed
27. Alameh N (2003) Chaining geographic information web-services. IEEE Internet Comput
28. Snell J, Tidwell D, Kulchenko P (2001) Programming web-services with SOAP. O'Reilly Publisher, Sebastopol
29. Yin J, Carswell JD (2011) Touch2Query enabled mobile devices: a case study using OpenStreetMap and iPhone. In: Web and wireless geographical information system (W2GIS 2011). Springer, Kyoto, Japan, pp 203–218
30. Mazzetti P, Nativi S, Caron J (2009) RESTful Implementation of geospatial services for Earth and Space Science applications. Int J Digital Earth 2(Suppl 1):40–61
31. Ballatore A, Wilson DC, Bertolotto M (2013) Survey of volunteered open geo-knowledge bases in the semantic web; Quality issues in the management of web information. ISRL 50. Springer, pp 93–120

Spatial Analysis of Peasants' Migrations in Russia/USSR in the First Quarter of the 20th Century

Leonid Borodkin

Abstract The role of migrations in Russian history is immense and cannot be underestimated. This paper is the first one attempting to analyze (using specialized aggregation statistical method and GIS) the macrostructure of the migration flows of Russia's rural population since the end of the 19th century and to the first quarter of the 20th century to create an integral spatial distribution of Russia's rural population inter-regional movement. For such enormous country as Russia it's necessary to use macro-analysis of the migration data. The second section of the paper gives short overview of the peasants' migrations in Russia and reveals the "bottlenecks" in the relevant research literature. The third section shows that these "bottlenecks" can be overcome using the materials of the 1926 All-Union Population Census. The method is applied for the aggregation of the migration flows network according to the data on all the 29 regions of the country (the fourth section); our NetAgg program permits to construct a macrostructure of the inter-regional migration, to reveal the groups of the main regions—the "consumers" and "suppliers" of rural migrants, to determine the intensities of the migration flows between the revealed homogeneous groups of regions. The fifth section gives interpretation of the obtained macro-structure of peasants' migration flows and the last one contains GIS maps to visualize the results of spatial structure of the migration flows.

Keywords Historical GIS · Thematic mapping · Historical texts · Migration flows

1 Introduction

Tendencies of development of modern science are determined, mainly, by profound interdisciplinary processes and informatization of scientific researches. Such processes have affected both technical sciences and humanities.

L. Borodkin (✉)
History Department, Lomonosov Moscow State University, 27-4, Lomonosov Prospect (Shuvalovsky Body), Moscow 119991, Russia
e-mail: lborodkin@mail.ru

© Springer International Publishing Switzerland 2015
V. Popovich et al. (eds.), *Information Fusion and Geographic Information Systems (IF&GIS' 2015)*, Lecture Notes in Geoinformation and Cartography,
DOI 10.1007/978-3-319-16667-4_2

Development of branches of research at the junction of informatics and other sciences, including historical, has led to origination of new scientific domain called "digital humanities" [1, 3, 4, 11, 12 and 15]. This domain allows to adapt modern informatics methods and algorithms, automatized data processing methods, for peculiarities of humanities.

Currently, new informatics field for humanities, particularly for historical sciences, is being actively created. It is connected, firstly, with wide access to new digital sources of historical data. Secondly, approaches, methods and technologies of solving historical tasks are changing. New methods of processing complex initial data are arising.

We can observe a trend of convergence of historical and geographical studies, and geographical information system (GIS) is one of the best apparatus for such integration. GIS allows to graphically represent geo-spatial information, retrieve related data, add and manage attribute data, execute analysis of given data and etc. For historical science GIS can become an integrated apparatus for simultaneous processing of heterogeneous types of information: attribute data, spatial data, time series of data and etc. GIS is also able to manage large sets of data, allows executing integrated spatial analysis methods and methods of processing quantitative and qualitative types of data. Application of GIS allows historians to solve problems, associated with analysis of spatial-temporal alterations of territory, based on various historical statistic data. Also it can give spatial representation for given initial information and research results as well. As data sources for GIS historical texts and cartographic materials can be used.

GIS, by no means does not apply to replace classical methods of historical study of history. Nevertheless, GIS can give enough of advantage to scientists by more efficiently managing all sorts of historical data available along with providing data graphic representation.

However, despite the advantages GIS can give to a scientist, its application in historical field of study is currently underdeveloped. This predicament can be solved by several approaches. Firstly, specific algorithms and automatized methods of data processing for historic researches should be developed. Secondly, scientists should respond more actively to emergence of innovative technologies and program products in geoinformation field and should adapt them for application in historical science.

2 Peasants' Migrations in Russia and Their Studies

The active work on the problem dates back to the 19th century. Nowadays an important place in this field belongs to the works dealing with (exclusively, or inclusively) the analysis of directions and values of the particular migration flows, to reveal their general regularities. The most stable of the regularities is the one dealing with the conception of A.A. Kaufmann, according to which the vagrant "instinct", one of the characteristic traits for the Russian peasant's character in all the times, displayed itself "one-sidedly": the main directions of the peasants' migrations went finally from the Russia's center to its outskirts.

The breakup in Russia's peasants' migration started just after the Great reform of 1861 abolishing serfdom. Beginning from the post-reform time a wave of migrations within the country's limits was continuously growing. A special explosion registered in the 1880s–1890s was related to the beginning of the active peasants' colonization of some regions of Siberia and Far East. The other impetus, even more powerful, was given by agrarian reform prepared by P.A. Stolypin at the beginning of the 20th century. During 20 years (1885–1904) 1.491 million peasants left the European part provinces, however in the next 10 years the number of such migrants reached 3.139 million (Sklyarov 1962).

On the whole, a wave-like pattern is peculiar to the spatial migration of Russia's population. The causes of its variation include the startup of the Trans-Siberian railroad, the cyclical pattern of the economic evolution, the explosions of social conflicts, the regional specifics of the agrarian relations, rumors, bad crops, wars, etc. And last, but not least, is the State's position in the migration problem; this position changed many times. The practice of the administrative suppression of peasants' migration to the "free border regions" was substituted by the policy directed at restraint, when possible, of these processes, that didn't, anyway, exclude the good old "well tested" methods in solving of these problems: even in this period the use of the police and even of the army was common in suppressing the migration movement.

The transition from restriction to stimulation of the migration of large strata of population to the far off borders dates back to the 6th of July 1904. That was the day when the new direction of the migration policy was legalized at a declared level. This improved the opportunities of anyone who desired to take part in the stimulated and even publicized by the Government migration process. The special Migration Department has been established under the Agriculture Ministry in that year. This body remained under the Soviet power, too; it became one of the departments of the People's Commissariat of Agriculture.

2.1 "Bottlenecks" in Historiography

Though there is a number of studies on the history of migrations both in the Czarist and the Soviet period there are practically no works dealing with the peasants' migrations within the USSR borders not "cut" (from below or from above) by the 1917 frontier. In other words, a tradition has formed to analyze the domestic migrations in Russia before and after 1917 as two different, completely independent processes. This is the first "bottleneck". As we have revealed, there exists a clearly expressed succession in the essence and the character of the main migration processes between Russia's pre-revolutionary epoch and the first post-revolutionary decade.

The traditionally centrifugal main directions of migrations remained in the early Soviet Russia as well. Legalizing de facto the existing realities the special edict of the USSR All-Union Central Executive Committee and Council of People's Commissars of July 6, 1925 declared the start of the planned migration to the Volga region, Siberia and Far East. The other decree completed this list with the Northern

Caucasus and Urals. The character of the State participation in the migration processes, up to the second half of 1920s stayed in principal unchanged. The Soviets, in accordance with the critics of the autocracy's colonial policy willing to correct the errors of their predecessors, returned in the first years of their rule to the policy of constraining the pressure of the migration wave. But this pressure turned out to be so high (though it never reached the former values) and the problems it raised so acute that the end of 1924—beginning of 1925 gave start to the transition to the "planned migration". The authors of this term were exclusively Soviet theoreticians and organizers of the migration policy, but its sense was quite different in the '20s and the '30s, when migrations within the USSR borders really lost their former sense under the highest level of regulation of the social life.

Analyzing the literature on the evolution of migration processes in Russia one has to state that the picture in a whole seems fragmentary. Characterizing the second "bottleneck" of the analytical historiography, it must be noted that the most widespread is the approach according to which any author chooses as the object of research some particular migration flow ending in a specific region or vice versa the flow originating from a specific region. There's no sense in negating the usefulness of these works. But the value of a particular subject can be fully seen only in the context of a whole background, the integrated structure of all the migration processes in Russia.

Finally, the third "bottleneck" is associated with the fact that very seldom an attempt to match "inputs" and "outputs" of migration flows is being made in works investigating the regions with the most intense out-migration or settlement of migrants from other regions. When "supplier" regions are matched with "consumer" regions, there is still a lack of proper differentiation of the migration flows structure that prevents us from "conjugating" the two corresponding regions [6]. On the other hand, it should be noted that the Soviet/Russian historiography has undertaken to construct migration "supplier-consumer" chains of regions to describe migration movements of Russia's rural population at the turn of the 20th century. However, all these attempts have not gone beyond the limits of specific regions to elucidate only particular fragments of a complicated network of migration routes, and reflected one or another researcher's concern with migration links of a particular region. In this respect of keen interest is the research field working out a concept of landscape similarities between the old and new homeland of settlers [17].

Thus, to sum up what was said above about the bottlenecks of the historiography, the following should be mentioned.

Despite a great number of works that analyze the various aspects of migrations in Russia over the period under consideration there is still no general picture of the migration flows structure giving a systematic presentation of the processes? The main causes are associated, to our mind, with the following historiographic traditions:

- isolated studies of migration processes in the period of the pre-revolutionary development and the first post-revolutionary decade;
- the dominance of local studies limited to the migration processes of a particular region;

- lack of the match between "inputs" and "outputs" of migration flows in the works that surpass the limits of local analysis.

To analyze the migration flows structure as a system and to produce an integral spatial distribution of Russia's rural population migration at the end of the 19th century and the first quarter of the 20th century it is essential to eliminate these bottlenecks and to use the relevant sources and analytic techniques.

What are the requirements that a source should meet to cope with this task?

It appears necessary for a source contain data on the migration flows intensities in every match of the country's regions. The data should be differentiated by principal social categories of migrants (including peasants as a separate category) and reflect the settlement-out-migration dynamics for each region over the period under consideration. The number of regions should be large enough and run into several dozens to represent the regional specificity in detail. The source of that kind is presented in the next section.

3 The All-Union Census of 1926: Potentialities and Limitations

Out of all the sources that describe the country's rural population migration flows from the end of the 19th century to the first quarter of the 20th century, the All-Union 1926 Population Census data satisfy the above-stated requirements to a great extent. The final results of the Census were published in 56 volumes which were divided into seven sections. Section III ("Marital Status. Birth-place and length of residence". Vol. 35–51) is of the most interest for the purpose of our study. The migration flows intensities data were obtained by summing up the answers to the questions 6 and 7 of the census personal questionnaire ("Where were you born: here or not? If not, specify your birth-place and how long you are domiciled in here"). Table VI of the Census Section III ("Natives of other regions by birth-places. The results by social groups") summarizes the data as a matrix that comprises the number of migrants for every match of regions. The number of regions totals 29, they correspond to the USSR regional division for the year 1927. The census recorded 3,605,314 peasant that were natives of regions other than they lived in.

Relevant information about migration flows is also offered by Table IV of the Census ("Population by length of residence, nationality, occupation and employment in branches of economy") and V ("Natives of regions, other that they are domiciled in birth-places and residence").

In fact, the Census data cover the territory of the whole country allowing us to overcome the limits of local research with "inputs" and "outputs" of each migration flow defined clearly by a corresponding match of regions. These data permit us to study particular migration flows for rural migrants without cutting flows at the time cross-section of October, 1917.

Every value a_{ij} in these matrices shows how many natives of a region specified in line i live in another region specified in column j. In the foreword to Section III of the Census the main objectives of the tables' data are described as follows: "(a) to study the directions of migration flows inside the country; (b) to reveal centers that attract population, and (c) to reveal the regions which population is drawn towards in these centers".

To what extent do the final data of the first All-Union Census meet the above-said requirements for a source on a rural population migration structure? This source, to our mind, enables us to eliminate the bottlenecks of historiography. In fact, the data of Table IV of the Census cover the territory of the whole country, allowing to overcome the limits of local research with "inputs" and "outputs" of each migration flow defined clearly by a corresponding match of regions. These data permit us to study peasant migration flows without cutting the flows at the time cross-section of October 1917. The data of Table IV of the Census also make it possible to study dynamics of the migration flows intensities for every "consumer" region. In this table the summary flow of settlers in each region is apportioned by nine time spans, five rather short spans fall in the period of 1917–1926, four lengthier ones cover the pre-revolutionary period.

4 Processing the Census Data

Let us proceed to a preliminary stage of the census data analysis, associated with calculations of most general parameters of inter-regional migration activity.

In this section we will consider indices, characterizing various aspects of the peasants' migration activity for each of 29 regions:

I—coefficient of migrants increase,
D—coefficient of migrants decrease,
B—coefficient of migration balance,

where

$$I = (V_i/V) \times 100\,\% \tag{1}$$

$$D = (V_d/V) \times 100\,\% \tag{2}$$

$$B = (V_i - V_d)/(V_i + V_d) \times 100\,\% \tag{3}$$

where V_i (or V_d) is a number of migrants that settled in a given region (or left it) over the time span under consideration, V is the total number of peasants at the moment of census.

While I and D describe the population migration mobility of a particular region, the coefficient B represents relative predominance of a settlement flow over an out-migration flow. Its value varies from $-100\,\%$ to $+100\,\%$, with plus and minis indicating which of two flows dominates.

Table 1 offers *I*, *D*, and *B* values of rural migrants for all the 29 regions of the country.

Highest values of the increase coefficient (*I*) indicate a substantial share of newcomers among the farmers in Siberia ($I = 32.7\%$), the Far East (23.3), Kazakhstan (15.5), Crimea (10.6) and Northern Caucasus (8.4). Whereas in central regions, as well as most regions of Ukraine and Byelorussia, this index has low values. Instead, it is precisely in Byelorussia and Ukraine where the decrease coefficient (*D*) is highest (11.7 % in Dnepropetrovsk region, 11.2 in the Left-bank Ukraine, 9.5 in Polessie, 8.9 in Byelorussia, 8.7 in Mining Industrial region).

Table 1 The principle parameters of the regional peasants' migration activities

	Regions	V	V_i	V_d	I (%)	D (%)	B (%)
1.	Northern	1,323,877	20,212	26,957	1.5	2.0	−14.3
2.	Leningrad-Karelian	2,617,849	24,598	136,558	0.9	5.2	−69.5
3.	Western	2,242,619	51,806	180,434	2.3	8.0	−55.4
4.	Central Industrial	8,363,813	121,074	255,273	1.4	3.1	−35.7
5.	Central Blaclcsoil	6,287,234	80,967	499,192	1.3	7.9	−72.1
6.	Vyatka	2,048,876	21,137	128,959	1.0	6.3	−71.8
7.	Ural	3,187,729	137,593	111,639	4.3	3.5	10.4
8.	Bashkiria	1,436,148	94,076	43,367	6.6	1.0	36.9
9.	Middle Volga	5,828,751	126,267	321,920	2.2	5.5	−43.6
10.	Lower Volga	2,542,530	76,499	107,570	3.0	4.2	−16.9
11.	Crimea	191,167	20,365	11,389	10.6	5.9	28.3
12.	Northern Caucasus	3,825,108	322,046	98,148	8.4	2.6	53.3
13.	Dagestan	311,609	13,015	4957	4.2	1.6	44.8
14.	Kazakhstan	3,011,978	466,070	46,023	15.5	1.5	82.0
15.	Kirghizia	465,409	36,261	10,114	7.8	2.2	56.4
16.	Siberia	4,258,495	1,359,215	55,784	32.7	1.3	92.1
17.	Buryat-Mongolia	255,855	17,706	6864	6.9	2.7	44.1
18.	Yakutia	166,439	1346	662	0.8	0.4	34.1
19.	Far East	676,302	157,290	12,297	23.3	1.8	85,5
20.	Byelorussia	2,706,481	32,283	242,229	1.2	8.9	−76.5
21.	Polessie	1,642,794	28,821	156,153	1.8	9.5	−68.8
22.	Right-bank Ukraine	4,930,372	50,990	282,547	1.0	5.7	−69.4
23.	Left-bank Ukraine	3,714,564	69,057	415,899	1.9	11.2	−71.5
24.	Steppe	2,860,976	147,650	222,653	5.2	7.8	−20.3
25.	Dnepropetrovsk	1,175,354	65,159	137,621	5.5	11.7	−35.7
26.	Mining Industrial	606,427	40,371	52,666	6.6	8.7	−13.2
27.	Transcaucasus	1,755,038	6767	25,015	0.4	1.4	−57,4
28.	Uzbekistan	1,796,042	12,823	9535	0.7	0.5	14.7
29.	Turkmenia	403,400	3853	2889	0.9	0.7	14.3

Source Tables IV, V All-Union Census of Population (1926) Vol. 35–51

Table 1 contains quantitative assessments of migration activity of the regions known in historiography as "consumers" and "suppliers" of rural migrants. The migration balance coefficient values confirm that maximum prevalence of settlement over out-migration of peasants was typical of the outlying regions of the country: Siberia ($B = 92.1\%$), the Far East (85.5), Kazakhstan (82.0), Kirghizia (56.4) and Northern Caucasus (53.3). On the contrary, maximum predominance of out-migration over settlement of farmers was recorded in Byelorussia ($B = -76.5\%$), Central Blacksoil region (-72.1%), Vyatka region (-71.8%), Left-bank Ukraine (-71.5%), and Leningrad-Karelian region (-69.5%).

Table 1, however, gives no evidence of the migration "mainstreams" structure. To make it possible we calculated the matrix S which contains values of a structural migration coefficient. S_{ij} is the number of peasants—natives of ith region who settled in jth region divided by the total number of peasants emigrated from ith region. This matrix provides data to construct the optimal migration typology of the country's territory.

The network structure grouping method is used here for the aggregation of the migration flows structure according to the data on all the 29 regions of the country. That permits us to construct a macrostructure of the interregional migration network; to reveal the groups of the main regions which are "consumers" and "suppliers" of migrants; to determine the intensities of the migration flows between the homogeneous groups of regions.

The aim of the method is to single out subsystems in the network system under consideration and to disclose the structure of links between the subsystems. Such a classification of migration flows permits us to proceed from hundred (if not thousands) of minor relations to a small number of major relations between the revealed subsystems. It is essential that the concise description of the migration flows structure is provided not by the selection of strongest links, but by its aggregation of the whole multitude of relationships. This approximation method of the weighed graphs aggregation (AMA), taking into account all the migration flows and discovering groups of homogeneous units, is realized both in a fuzzy and "hard" versions [5]. The properties of the quasi-optimal solution of the aggregation problem were proved, and an algorithm to minimize the value of the functional proposed to estimate the quality of approximation was constructed. The program NetAgg [16] created to realize AMA approach contains the procedure of a discrete optimization as a principle block.

5 Macrostructure of the Interregional Peasant Migration Flows

Analysis of the migration network aggregated structure revealed the optimal number of homogeneous groups (macro-regions) to be equal to eight. Aggregation of the S matrix realized on the basis of AMA method resulted in the following composition of the homogeneous macro-regions:

Table 2 Aggregated structure of interregional peasants' migration flows: Matrix S of the outmigration structural coefficients (%)

Region number								
	I	II	III	IV	V	VI	VII	VIII
I	–	31.8	5.2	1.5	0.5	0.1	9.5	2.1
II	19.0	–	8.9	1.6	7.3	0.1	0.8	3.6
III	10.6	27.5	–	2.7	1.3	5.1	1.8	2.0
IV	27.5	14.3	12.3	3.5	0.3	0.3	1.7	1.3
V	1.5	42.4	3.0	0.8	14.0	1.3	0.2	1.7
VI	1.2	1.6	62.1	1.1	0.4	11.4	0.1	1.3
VII	40.9	2.4	2.5	0.8	0.1	0.0	16.6	1.5
VIII	45.4	9.0	4.1	0.7	0.2	0.1	0.8	4.0

I   Siberia;
II   Kazakhstan;
III Northern Caucasus;
IV   Central Blacksoil, Crimea, Dnepropetrovsk, Mining Industrial, Right-bank Ukraine, Left-bank Ukraine, Byelorussia, Polessie, Steppe;
V   Kirghizia, Turkmenia, Uzbekistan;
VI   Dagestan, Transcaucasia;
VII Buryat-Mongolia, Far East, Yakutia;
VIII Central Industrial, Northern, Leningrad-Karelia, Western, Vyatka, Ural, Bashkiria, Middle Volga, Lower Volga.

One can notice that each of these eight groups consists of adjacent territorial units. It is the reason to use the term 'macro-region' in this case. The first three regions have very special position in this aggregated structure, taking into consideration their composition (each of these three groups consists of one territorial unit) and their extremely important role in the peasant migration flow: 59.6 % of all migrants settled just in these three regions. This role is reflected in Table 2, which shows the structure of migration flows between the eight groups.[1]

Each figure in Table 2 is a value of the out-migration structural coefficient S_{ij} calculated for the corresponding pair of aggregated macro-regions. The graphic representation of this aggregated structure is given by Fig. 1, constructed from

---

[1] Each figure in table is a value of the outmigration structural coefficient S_{ij} calculated for the corresponding pair of aggregated macro-regions. For example, Fig. 62.1 in the 6th line and in the 3rd column indicates the average proportion of peasants-natives of Dagestan and Transcaucasia, settled in the North Caucasus, which was equal to 62.1 % from the total number of peasants emigrated from Dagestan and Transcaucasia.

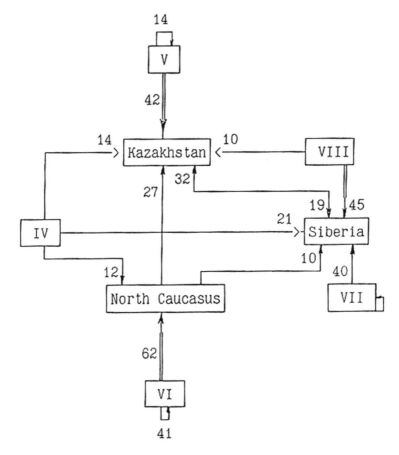

Fig. 1 Macrostructure of peasants' migration flows (the end of 19th—the first quarter of 20th centuries). Numbers have the same meaning as those in Table 2

Table 2. Each edge of the graph coordinates to significant migration flow (the threshold value of the matrix element in Table 2 is 10 %).

Figure 1 gives the evidence of the 'attractors' in the peasant migration structure. These are Siberia, Kazakhstan and Northern Caucasus. It is worth to notice that significant out-migration flows from these three regions were also directed to one (or two) of that regions. Figure 1 shows the migration flows between other five macro-regions, which were basically 'suppliers' of peasant migrants.

NetAgg software made it possible to construct the aggregated spatial picture of the peasant migration in Russia/USSR in the first quarter of the 20th century.

Fig. 2 Spatial distribution of peasants migrations in the end of 19th—beginning of 20th centuries. based on net migration coefficient B (see formula (1)

6 The Spatial Structure of Migrations: Using GIS

Historical cartography is based nowadays mainly on application of GIS technology. Since 1990 a number of works were published in this area (see for instance [7–9, 13 and14]).

In this study, we used GIS MapInfo PRO. The creation of digital cartographic ground was carried out on the basis of the map of Soviet economic regions produced by the USSR Central Statistical Agency on 23/08/1927. We formed GIS layers and linked them with statistical data on interregional migration flows. These layers include the boundaries of 29 economic regions of the country, eight macro-regions (obtained by statistical aggregation of the interregional migrations matrix), cities, migration flows, islands, grid, frame. Creation of this GIS map implies application of both mathematical basis (coordinate system—SC, map projection, scale) and vector conversion according to the various sources.

Our GIS maps show net migration coefficients for each of 29 economic regions of the USSR (Fig. 2), as well as the directions and intensities of the main migration flows between eight macro-regions (Fig. 4). Figures 3 and 5 represent the same spatial distributions for migration flows of hired agricultural workers.

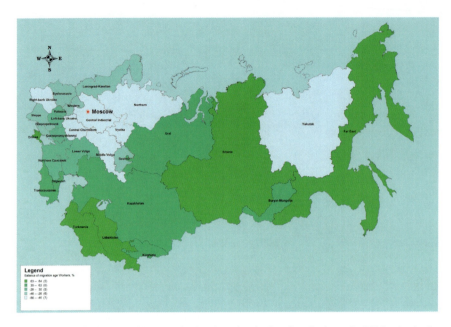

Fig. 3 Spatial distribution of hired agricultural workers' migrations in the end of 19th—beginning of 20th centuries based on the net migration coefficient B (see formula (1))

Fig. 4 Spatial distribution of peasants migration flows in the end of 19th—beginning of 20th centuries based on the structural migration coefficient S_{ij}

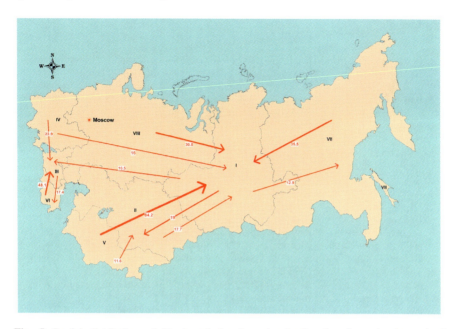

Fig. 5 Spatial distribution of hired agricultural workers' migration flows in the end of 19th–beginning of 20th centuries based on the structural migration coefficient S_{ij}

7 Conclusion

Currently, main trend of application of geoinformation in historical researches is application of digital cartography. Involvement of new sources of data in the process of historical research allows to achieve new knowledge about the process under consideration.

Methodology of peasants' migration flows analysis, executed with standard program products as NetAgg and MapInfo, is described in this paper. Application of computing aids allowed to process larger set of original sources, containing information about peasants' migration processes, in the course of research.

However, we should note a certain disbalance between arising of new methods and technologies of processing large sets of data and their application in practice. Thus, there is a need for development of complex method of geoinformation integration of geoinformation in historical science. It is evident that to solve this task, close collaboration between scientists of both scientific domais is required.

References

1. Berry DM (2012) Understanding digital humanities. Palgrave Macmillan, Basingstoke
2. Beveridge AA. Immigration, ethnicity, and race in metropolitan New York, 1900–2000. Past time, past place, pp. 65–77.
3. Bod R (2013) A new history of the humanities: the search for principles and patterns from antiquity to the present. Oxford University Press, Oxford
4. Borgmann C (2009) The Digital Future is now: a call to action for the humanities. Digital Humaniti Q 3(4)
5. Borodkin LI (1979) Aggregated structure of graphs with fuzzy blocks. Autom Remote Control 29(8):142–153
6. Bruk SI, Kabuzan VM (1980) Dinamika i etnicheskii sostav naseleniia Rossii v epokhu imperializma (konets XIX veka 1917). Istoriia. SSSR 3:77–79
7. Ell Paul S, Gregory Ian N (2001) Adding a new dimension to historical research with GIS. Hist Comput 13(1):1–6
8. Gregory IN (2003) A place in history: a guide to using GIS in historical research (AHDS guides to good practice). Oxbow Books Ltd, Oxford
9. Gregory IN (2008) Historical GIS: technologies, methodologies, and scholarship (Cambridge studies in historical geography). Cambridge University Press, New york
10. Gregory IN, Kemp KK, Mostern R (2001) Geographical information and historical research: current progress and future directions. Hist Comput 13(1)
11. Hillier A. Redlining in Philadelphia. Past time, past place, pp 79–92
12. Hockney S (2004) The history of humanities computing. In: Schreibman S, Siemens R, Unsworth J (eds) Companion to digital humanities. Blackwell, Oxford
13. Knowles AK, Hillier A (2008) Placing history: how maps, spatial data, and GIS are changing historical scholarship. ESRI, Inc, Redlands
14. Knowles AK (2002) Past time, past place: GIS for history. ESRI Press, Redlands
15. Svensson P (2009) Humanities computing as digital humanities. Digital Humaniti Q 3(3)
16. The NaaS NetAgg system. URL: http://www.harness-project.eu/
17. Yakimenko NA (1980) Sovietskaia istoriografiia pereseleniia krest'yan v Sibir'i na Dal'nii Vostok (1861–1917). Istoriia SSSR 5:91–104

Part II
Web-Based and Location-Based GIS Applications

RRM—A Referenced Routing Model to Generate a Semantic Service of Navigation in Mobile Devices

Ana María Magdalena Saldaña-Pérez, Miguel Torres-Ruiz, Marco Moreno-Ibarra and Giovanni Guzmán-Lugo

Abstract With the inclusion of the Global Positioning System (GPS) on mobile devices, users have a new option to obtain routes and travel information. Most of the routing systems advances are focused on time improvements; actually, the problem with commercial routing services is the way in which they give their route instructions, since they do not include support features that could be useful as landmarks or points of interest. By analyzing these issues, this paper proposes the Referenced Routing Model (RRM), which provides to the users semantic instructions to go from one place to another, by using geospatial analysis tools, semantic processing and mobile technologies. The obtained routes begin at the mobile position and show to the user the points of interest and visual references located on the road in order to aid him during his trip. An ontological model has been designed in order to provide the references related to the routes.

Keywords Ontology · Point-of-interest · Routing · Geospatial analysis · Global positioning system

---

A.M.M. Saldaña-Pérez (✉) · M. Torres-Ruiz · M. Moreno-Ibarra · G. Guzmán-Lugo
Centro de Investigación en Computación, Instituto Politécnico Nacional,
Mexico City, Mexico
e-mail: asaldana_a12@sagitario.cic.ipn.mx

M. Torres-Ruiz
e-mail: mtorres@cic.ipn.mx

M. Moreno-Ibarra
e-mail: marcomoreno@cic.ipn.mx

G. Guzmán-Lugo
e-mail: jguzmanl@cic.ipn.mx

© Springer International Publishing Switzerland 2015
V. Popovich et al. (eds.), *Information Fusion and Geographic Information Systems (IF&GIS' 2015)*, Lecture Notes in Geoinformation and Cartography,
DOI 10.1007/978-3-319-16667-4_3

1 Introduction

The advances in GPS technologies are focused on the routing service. In spite of having increased the user possibilities to obtain good results for route searches, GPS technologies are interested on time and processing improvements, do not into having a better communication with people [12].

Thus, go to a specific place in a city could be a difficult situation, especially if either the area or precise ending point is unknown; sometimes the routes instructions are not clear. For example, imagine a person who wants to go from his office to the nearest mall center; when he asks to his GPS navigator for instructions, the result could be: "walk 250 m to the left, walk straight ahead 100 m, turn to the right, walk straight ahead 85 m, turn to the west 20 m, walk straight ahead 100 m". Sometimes for a pedestrian, it is difficult to have a good orientation and a real perception of distances; that is the reason why some users feel unsatisfied with the instructions given by the routing services. The main goal of our model is to provide semantic instructions that users can cognitively understand in such a way as human beings perceive real world.

According to above, we propose the Referenced Routing Model (RRM), which is based on routing with Points-of-Interest (POIs). It combines geospatial analysis tools and an application ontology in order to semantically describe places such as shops, schools, hospitals, etc., like instances of the ontological model. In addition, web and mobile technologies are used to compute a route with nearby sites of interest. The instructions describe the direction changes and the POIs located on the route; such POIs were previously obtained from the ontology. The RRM is compound by four stages, recognition, routing, semantic processing and filtering. The *Recognition* stage determines the beginning and the ending of the route by using a mobile application implemented in Android. The data are used in the *Routing* stage in order to generate the path by the usage of the *A star Pgrouting* algorithm, taking into account the study area's roadways. The *Semantic processing* stage is responsible for searching landmarks semantically from the application ontology designed for the RRM. The *Filtering stage* consists on selecting some business according to the user's interest in order to show them on the route.

As case of study, we have analyzed a fragment of *Gustavo A. Madero District*, at Mexico City. The proposed project generates a descriptive route that shows the nearby business to the segments that are part of the path, by searching into the ontology the name of the streets, the business located on them and their classification. The data used to generate the routes belong to the *Gustavo A. Madero District*, and the information about business and POIs was taken from the Mexican National Statistical Directory of Economical Units (DENUE).

The paper is organized as follows: Sect. 2 describes the related work studied for this investigation. Section 3 describes the component stages of the RRM methodology. Section 4 depicts the results obtained by applying the RRM methodology as well as the comparison between the RRM data processing and some other commercial routing services. Finally, Sect. 5 outlines the conclusion and future work.

2 Related Work

The GPS routers on mobile devices have been improved by the implementation of GIS tools, which increases their quality and functionalities; furthermore, production and acquisition costs have decreased due to people demand [3, 9]. The problem with the GPS routers is that sometimes they are used at places where the accuracy on data is not good, as consequence, the information to build routes does not have good quality in order to produce meticulous results as in some other areas. Other case to consider is the kind of words used by GPS routers to notify changes of direction, since they are longitudinal measures and orientation terms; all the mentioned facts derive in problems for the user to understand the instructions [15], furthermore a GPS accuracy depends on the routing algorithm that applies, and the way that it obtains and processes data [10, 14]; this algorithm should be adaptable to the communication network, taking into account a wide background of information. The algorithm proposed by Jianhe Du and Lisa Aultman [7] accomplishes with those characteristics. This algorithm is one of the most specialized on the routing analysis because it studies the routes used by a group of users, considering each person activities and the time spent on them, the routes used by individual subjects, and routes used by people that share a vehicle. In [7] an algorithm that helps people to improve their routes, and notify them if there is a mistake in their route is proposed. It considers the applied changes to the route by the users or if they have spent more time than usual.

Up-to-date, the implementation of routing algorithms on mobile devices as smartphones is a challenge, because it is necessary to consider the operating system for each device. In despite of the portability and accessibility of smartphones, their battery holds its charge for few hours and they have an expected time of data acquisition of 10 s, they are slower and less accurate than dedicated GPS devices [16]. On the other hand, most of the advances on routing are focused on improving time and path's length.

In order to define an object as POI, it is needed to consider the relevance of the place represented by the object, if it can be easily identified, and its visual, structural and semantic properties [15]. Ontologies are very useful to describe real information in order to employ it into specific systems; its use on routing systems makes easier to relate geographic features with their attributes, in order to handle and understand them. Furthermore, ontologies can be used to rank concepts by their importance and to establish relations between them [2]. Due to their capability to simulate a real environment and create relations between concepts, ontologies have grown-up in many areas, such as economy, medical services, routing, among others [11]. Particularly on routing projects, the ontology provides to the system, information about geographical features, and aids them to automatize the decisions making by stablishing relations between the domain and the concepts. The ontologies usage by routing systems has its origins on Kulik work [6].

There are some commercial routing services for mobile devices, one of them is Google Maps for Android [4], this service shows users their actual position and the route to arrive to a specific place selected by the user. The instructions to follow the route can be directed by voice or text; it shows the remaining time to reach the objective and real-time traffic. Google Maps for Android allows users to share their opinion about a place, and photos in its social network, users can visualize three dimensions maps, street views and interior building maps. The problem with this service is that user needs to be connected all time; otherwise he could not request information or visualize its route.

Nokia Maps, known commercially as Here Maps, is a service for Nokia cell phones and mobile devices; it provides cartography to devices with Symbian PS, Windows Phone 7, Firefox, among others. Nokia Maps has an application to give voice instructions to pedestrians or drivers. Nokia Maps lets user know the traffic situation in some cities, and share their coordinates by using social networks [5]. Waze is one of the most used services to obtain information about traffic and routes with less vehicular problems. Waze has built a user community in order to compile information, where each user is able to notify vehicular problems and show his position by the continuous sensing of the mobile device. It obtains real-time data and alternative routing [13].

3 The RRM Methodology

As a new implementation for GPS routing systems, we have designed a method to include POI references on the route and semantic instructions; these references are buildings or business located nearby the route. Such trades are chosen to aid to the user during his trip and they are obtained by a semantic process applied to the ontological model. In addition, we improve a new feature to the methodology that allows user to decide if he wants or not to visualize a specific kind of business neighboring to his route. The methodology is compound of four stages: recognition, routing, semantic searching, and filtering. The expected outputs for this work are: a route that connects the user with the ending place of his interest, a map with the business and landmarks located on the route that are used as references on the given semantic instructions, and the extra business, which commercial activity has been selected by the user. In Fig. 1, the general framework of the RMM methodology is described.

In the *Recognition* stage, the system obtains the coordinates of the GPS device (beginning point), and the name of the place where user wants to arrive (ending point), by using the RRM application that was implemented on Android [1]; such application consist mainly of two interfaces: the first one designed to obtain user information, and the second one, that presents the obtained results. On *Routing* stage, the system computes the shortest route between the beginning and ending points, by using the *A star* algorithm directly from *Pgrouting* [9]; some spatial functions, and the road network of the study area are considered in the calculus. To process the network road and the user coordinates, a spatial database has been created, it contains the roads of the area and it has been imported to the database on

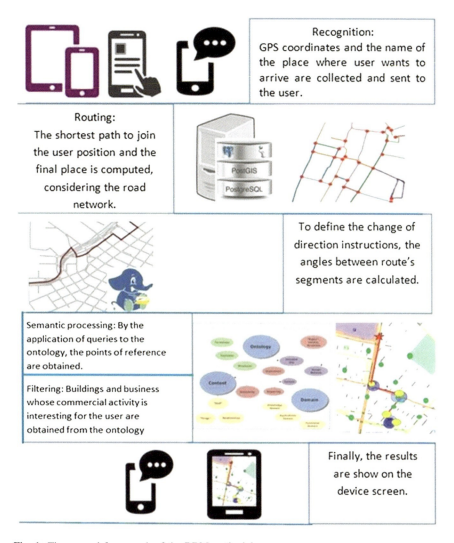

Fig. 1 The general framework of the RRM methodology

PostgreSQL by using its spatial extension PostGIS. Once the table has been created, RRM works with its data, applying queries and spatial functions by using a SQL editor of PostgreSQL. Figure 2 shows a diagram that relates the network road with its table representation.

The *Semantic processing* stage uses the application ontology to obtain the representative buildings and business located on the route segments. In order to build the ontology a Methontology approach was used [11]. By considering the classification of roads in Mexico, we have identified four main classes: primary, secondary, tertiary and pedestrian roads. Each class is divided into entities that

Fig. 2 Relation between road table data and road network diagram

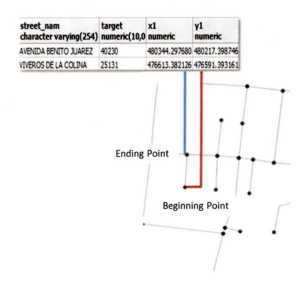

Table 1 Properties of the ontological entities

Ontology element	Properties	Sub-properties
Entity	Name	
	Traffic	
	Cost	
	Business	Name of business
		Commercial activity
		Business coordinates

describe a real road object with its individual properties, presented in Table 1. To retrieve information about a certain street or road, it is necessary to apply queries on the table generated by Post-GIS. With this information the system obtains the name of the streets that are part of the route, and the name of the business located on them, in order to give references to the user.

The *semantic search* is related to the last process, named *filtering stage*, where the application ontology is used to consider the user preferences, extra features such as shops or bus stations are added to the map. The SPARQL query applied to the ontology is executed on the mobile device. The ontology has been implemented in OWL and edited in Protégé [12]. The query requests to the ontology as well as the business located on the defined road are retrieved using the ontology's semantic reasoner. In this case, the name of the street that is part of the route calculated by Pgrouting is received, once the street has been found, the ontology returns to the mobile application, the instance related to the name of the road requested, its classification, the business located on it and the coordinates of each building. The structure of the data sent to the ontology is shown in Eq. 1.

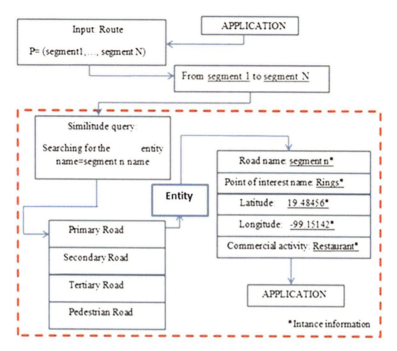

Fig. 3 Query procedure to obtain information from the ontology

$$F(P) = (route\ (segment1, segment2, segment3, \ldots, segmentN)) \quad (1)$$

Each route segment is searched in the ontology by comparing the segment's name with the names of the roads saved into the ontology that belong to the same class.

When a matching is found, the ontology sends the street attributes to the application. Figure 3 depicts the query process to obtain information from the ontology model.

The code to access to the ontological model on the mobile device has been implemented in Android with the Jena framework. When the system has obtained information about the streets of the route, business, points of reference, and the route is represented on the map. Furthermore at each point that represents a direction change, a special icon that joins the point and the given instructions written on the top of the screen, has been drawn as a reference for users.

3.1 Instructions to the User

To notify user the instructions of direction change, the Android application has a text box on the top of the screen. Instructions just mention the direction that must be

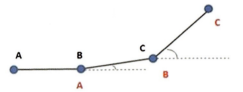

Fig. 4 Assignation of identifiers to calculate angles

taken and the business or important buildings located at that point, in order to avoid user misunderstandings.

The segments that compose a route are not totally linear, there are angles between them. To decide if it is needed to notify a direction change on the instructions, the angles between segments are computed trigonometrically, by identifying the beginning and ending of each segment, and the point where they are attached. According to the diagram on Fig. 4, the point labeled with the black A is the beginning of the first segment, the point labeled C is the final point of the second segment, and the point labeled with the black B, is the common point between segments.

To compute angles, it is necessary to deduct A latitude coordinates from B latitude coordinates, and take A longitude coordinates from B longitude coordinates. Subtractions are shown in Eqs. 2 and 3.

$$Ry = coordBx - coordAx \qquad (2)$$

$$Rx = coordBy - coordAy \qquad (3)$$

The angle between A and B points results by applying Eq. 4.

$$Angel\ 1 = at(Ry/Rx) = \tan-1(Ry/Rx) \qquad (4)$$

To obtain the angle between the C and A points, the arctangent of A coordinates from C coordinates deduction is done, applying the Eqs. 2–4. To obtain the final angle, it is necessary to subtract the angle formed between A and C, from the angle between A and B.

Once the angle between segments one and two has been computed, the reference points A, B and C are recalculated, now the point of joint between segment one and two is considered as the beginning point of segment two. The point C of the previous array is considered as the joining point between segments two and three; after the renaming of points, the equations are applied, until the algorithm has reached to the last segment of the route. The angle between the segments is compared with a defined interval [−14, 14] for determining if the change direction must be taken to the right or to the left. This interval has been computed by a probabilistic study applied on the road network. By comparing routes generated without a restriction against real routes; we noticed that in the road network design some segments should be straight. However, they have a little variation of 14 grades

Fig. 5 Assignment of direction according to the angle measure

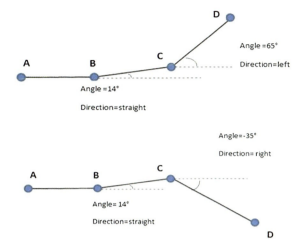

or less. So, in order to be sure that a change of direction must be taken, the interval to identify a real change was defined. It is clear that, if the system is applied to some other area with a different cartography it could be necessary to determine the minimum angle measure that represents a real change of direction.

In Fig. 5, directions and its relation with angles are depicted.

When a change of direction has been identified, the coordinates of the point B of that angle are stored, and a new query is done to the ontology in order to know the buildings on the street that have the same coordinates as the point stored, or those that are located near to it, with the objective to draw a yellow balloon to represent that point on the map and write the name of the business or building on the instructions text box.

In cases where the route includes a roundabout, RRM indicates to the user the output in which he has to change his direction, considering as "first output" to the output nearest the street where the user is walking. This is possible due to the geospatial process applied to the road network, in which all the road segments have a unique id. The sentence used by RRM to denote a change of direction in a roundabout is similarly to the next: "On the roundabout take the first output".

4 RRM Results and Comparison

4.1 The Mobile Application

The mobile application has been implemented on Android [1]; it can be used in any device with this operating system and GPS service. It allows the system to send data between user and server. It is composed of two interfaces: the first one was programmed for data acquisition and the second for visualizing results (see Fig. 6).

Fig. 6 Interfaces of the mobile application

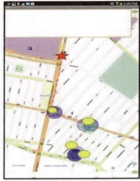

In the acquisition interface, the system calls the GPS device to obtain the user actual position and the name of the place that he wants to visit. In the routing stage the name of the place is matched with its coordinates stored on the server; after that, the coordinates are transformed into points to be displayed on the final map. The coordinates and its previous transformation into a point are developed on the PostgreSQL database. If the GPS device is not working, the coordinates of the user are obtained by an Internet connection. In the results interface, a map is displayed on the mobile device; the cartography used belongs to Open Street Maps [8]. On the map, the route is represented by a dashed line, the points of interest are represented by green and blue circles, the instructions of direction change are written on the top of the screen, and the references described on the instructions are symbolized on the map with a yellow balloon icon. The Android application receives from the ontology, the coordinates of the route segments and the coordinates of the business, by a *json* data structure that is sent from the server, these coordinates are converted into screen coordinates in order to draw the points on the map. If the connection is lost, the system does not lose the route and its elements, so user can visualize the final map and the results generated for the last query successfully made.

4.2 Experimental Results

To test the RRM methodology, it is necessary to install the RRM application on the device and store the ontology model on its memory. Let suppose a user that wants to visualize the route for going from his actual position to a church; when user initializes the RRM and writes the word "church" on the field "Place", the application shows to him the name of the nearest church to his position. After that, user has to decide the kind of business that is interesting for him. Green circles on the route will represent these features. When user accepts the ending place suggested by the RRM, the data are sent to the server by touching the button "Send". In Fig. 7, the acquisition interface and user's data are shown. In this case, the user has decided to visualize business and shops, whose commercial activity has a relation

Fig. 7 Acquisition interface with user information

with "medicine". When data are sent to the server, a table for this request is generated. The coordinates of the device are transformed into a geometric point, while the coordinates of the place that has been selected are searched on the server, and transformed into a point too. With this information RRM intersects the beginning and ending points obtained with the road network, in order to apply the *A star* algorithm and obtain the route for joining the points.

Once the routing is finished, the angles between segments of the route are computed, and the business and buildings are retrieved from the ontology. In order to obtain the trades, whose commercial activity is related to the user interest, a new SPARQL query is applied to the ontology, it returns those commercial features to be drawn by green circles on the map.

Finally, the instructions are written on the top of the screen, describing the direction changes needed to arrive to the place, and the names of business and POIs located at each change. As it is shown in Fig. 8, the points of reference are represented by blue circles, the points that represent business, shops and places whose commercial activity was selected by the user on the filter are represented by green circles. For this example, the green circles represent pharmacies and clinics, because user selected "medicine" as business filter. The route is indicated by a dashed red line, and the yellow balloon represents the buildings and business located on the direction change. If user needs to recalculate his route or he has decided to make a new request to the RRM system, it is necessary to touch the green arrow located on the bottom of the screen, this action will reinitialize the completely process.

All the time, the system is monitoring the user coordinates, in order to give the instructions one by one to the user while he is walking. In case of GPS bad

Fig. 8 Results obtained by the RRM methodology

performance, user can touch the icon of wireless Internet located on the bottom of the screen; by this way the system obtains the user coordinates through an approach. If the user is walking on the route but he wants to visualize on the map the location of a grocery store near to his path, it is only necessary to touch the green arrow, and select the word "grocery stores" on the "business filter" option; thus, these kind of features are drawn on the map as green circles, (see Fig. 9).

4.3 Results Comparison

In this paper we have studied the problem of users to understand GPS routes. The RRM methodology to provide routing instructions with POIs to users is described. With the purpose of test the RRM accuracy, we have compared the RRM routing results with some GPS mobile services. When we request to Google Maps for Android, a route to go from one place to another in our case of study area, we noticed that Google Maps gives us the shortest path for driving persons. However, if we ask for a pedestrian route, it gives us a different route, but sometimes it is not the shortest one for a walking person; the service of Google does not show to the user some reference points, in order to assure him to be in the correct road. The instructions of Google Maps are not very clear, because it uses an arrow on the top of the screen to show the way that the person should be walking (straight ahead or turn to a direction symbolized by the arrow), and the meters to arrive to the next direction change; so, if the person has not computed the real distance, he could commit a mistake and change his route. The advantage of Google Maps is that all the time, the application is receiving information from the server, so the route can

Fig. 9 Results considering the "grocery stores" selection filter

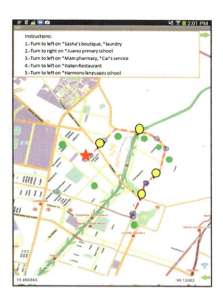

be recalculated when user changes his trip, and this monitoring makes possible to compute the remaining time to arrive to the ending point. The main problem of Google Maps for Android, is that in case of connection lost, it is not possible to visualize the map and the route, that is the reason why the device must be connected all the time. Another comparison was applied with Nokia Maps, this service is available for web and mobile Nokia devices, as any GPS service, it computes the route to go from one place to another, and shows the result on a map. The problem with Nokia Maps, is that the route only shows the beginning and ending points of the route and the name of the streets on the map, without any point of reference to help to the user. The instructions are accurate but unpractical for a walking person, because he cannot compute the long distances given by Nokia, and the service uses cardinal orientation terms resulting misunderstood for a non-familiarized person.

Finally, our system always shows the shortest path for a pedestrian user, the points of reference such as schools, hospitals, shops, restaurants and some other business and important buildings are depicted in order to aid the user interpretation of instructions, and avoid mistakes during the trip. The instructions do not describe measures of distance and cardinal orientation terms, they indicate the direction change to be taken and the name of a reference located there. By this way, the user can see the point of reference and change his direction with security. An extra component of our methodology is the "Business filter"; it is a tool for selecting the business that are interesting to visualize on the map, the RRM shows those elements located on the route or near to it, in order to satisfy a user specific requirement. This tool combines the ontology model with spatial information to provide more references to the user, and to know the position of some interesting places for him. In case of lost connection, the map continues available on the application with the last result generated by the system, so user can exploit this information in a critical

Fig. 10 Comparison between different routing services and RRM

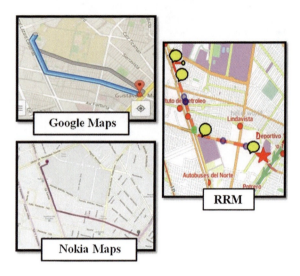

situation. The system could be improved in order to work at any city once the road network has been obtained.

Figure 10 presents the map and some requested routes in the three different GPS routing services analyzed, the time of response required by each routing service was measured on a device *LG Optimus L5*, and the same route was required to them. Google Maps was the fastest one with a response time of 12 s, Nokia Maps took 18 s, and RRM service had 17 s as response time. The RRM needs a little more time due to the SPARQL query, because it extracts POIs with a specific commercial activity, the request to the ontology for obtaining references, and its later representation on map. By taking into account the extra elements processing, RRM has a quick performance.

5 Conclusion and Future Work

Nowadays mobile devices are used in a lot of human activities; one of those is the displacement of people from one place to other, giving them instructions and routes. Most of the services for GPS routing compute routes for driving persons and give them instructions by using measure distances and cardinal orientation terms. Both features are very difficult to compute and understand for a person that is walking or driving; the new advances on this kind of routing services have improved the time and memory consuming, they are focused on technological issues.

We analyzed the user requirements in GPS routing services and formulate a solution for missing references and clear instructions proposed on the RRM methodology, a combination of geospatial processes and an ontological model to produce routes with POIs and references, to facilitate the instructions interpretation and avoid user mistakes during the trip. To model the application ontology, we used the road information given by the Mexican government; the ontology contains information about roads and POIs.

After the RRM process has finished, the user obtains a map with a variety of features for helping him to understand the given instructions and to let him know the position of places such as taxi stations, medicine shops, parks and commercial malls. In addition, the RRM system monitors the user position to recalculate the route if it is necessary. The RRM has been designed for pedestrian people, in case of lost connection the map and information are visualized without problems. The features shown on the route interact like auxiliary features in order to avoid user mistakes through his trip, and the study area can be updated or increased for covering other areas. The system is easy to use and works in any mobile device with GPS integrated and Android operating system, but it can be reprogrammed to work with other operating systems.

Acknowledgments This work was partially sponsored by the IPN, CONACYT and SIP, under grants 20140545 and 20140504. Additionally, we are thankful to the reviewers for their invaluable and constructive feedback to improve the quality of the paper.

References

1. Android (2014) Android mobile platform. http://www.android.com. Accesed 25 Jun 2014
2. Corcho O, Fernández-López M, Gómez-Pérez A (2002) Methodologies, tools and languages for building ontologies. Where is their meeting point? Informatics Faculty Universidad Politécnica de Madrid, Madrid
3. Czerniak RJ (2002) Collecting processing and integrating GPS data into GIS, National Cooperative Highway Research Program (NCHRP). Synthesis report 301 transportation research board, National research council, Washington, DC
4. Google (2014) Google maps. http://developers.google.com/maps/?hl=es. Accesed 25 Aug 2014
5. Here Maps (2013). http://m.here.com. Accesed 5 Feb 2014
6. Kulik L, Duckham M, Egenhofer MJ (2005) Ontology-driven map generalization. J Vis Lang Comput 16(3):245–267
7. London B, Huang B, Taskar B, Getoor L (2013) Collective stability in structured prediction: generalization from one example. In: Proceedings of the 30th international conference on machine learning, Atlanta, Georgia, USA
8. OpenStreetMap (2014). http://www.openstreetmap.org. Accesed 12 Jun 2014
9. Satyanarayanan M, LaMarca A, De Lara E (2008) Location systems: an introduction to the technology behind location awareness. Morgan & Claypool Publishers, San Rafael, p 88. ISBN:9781-59829-581-8
10. Seet BC, Liu G, Lee B-S, Foh C-H, Wong K-J, Lee K-K (2004) A-STAR: a mobile ad hoc routing strategy for metropolis vehicular communications. In: Networking technologies, services, and protocols; performance of computer and communication networks. Springer, Berlin, pp 989–999
11. Suárez MC, García R, Villazón B, Gómez Pérez A (2011) Essentials. In: Ontology engineering methodologies, languages, and tools. Ontological engineering state of the art. Ontology Engineering Group, Universidad Politécnica de Madrid, Madrid
12. Waluyo AB, Srinivasan B, Taniar D (2008) Research in mobile database query optimization and processing. Mob Inf Syst 1(4):225–252

13. Waze (2014) Waze livemap. http://es.waze.com/. Accesed 10 May 2014
14. Winter S, Truelove M (2011) Talking about place where it matters. Department of infrastructure engineering, Universidad de Melbourne. Lecture notes in geoinformation and cartography. Springer, Berlin
15. Wu Y, Winter S (2011) Interpreting destination descriptions in a cognitive way. Department of infrastructure engineering, University of Melbourne, Parkville
16. Xu C, Ji M, Chen W, Zhang Z (2010) Identifying travel mode from GPS trajectories through fuzzy pattern recognition. In: Seventh international conference on fuzzy systems and knowledge discovery (FSKD 2010). Key lab of Geographic Information Science, Ministry of Education and East China Normal University, Shangai, China

Semantic Trajectories: A Survey from Modeling to Application

Basma H. Albanna, Ibrahim F. Moawad, Sherin M. Moussa and Mahmoud A. Sakr

Abstract Trajectory data analysis has recently become an active research area. This is due to the large availability of mobile tracking sensors, such as GPS-enabled smart phones. However, those GPS trackers only provide raw trajectories (x, y, t), ignoring information about the activity, transportation mode, etc. This information can contribute in producing significant knowledge about movements, which transforms raw trajectories into semantic trajectories. Therefore, research lately has focused on semantic trajectories; their representation, construction, and applications. This paper investigates the current studies on semantic trajectories so far. We propose a new classification schema for the research efforts in semantic trajectory construction and applications. The proposed classification schema includes three main classes: semantic trajectory modeling, computation, and applications. Besides, we discuss the current research gaps found in this research area.

Keywords Semantic trajectories · Activity recognition · Data modeling · Data segmentation · Semantic applications · Sensor data

1 Introduction

Around 80 % of all the available data have either an explicit or an implicit geographical reference [1]. Explicit references are the actual geometries e.g., city boundaries, lakes, whereas implicit references are textual references to geographical

---

B.H. Albanna (✉) · I.F. Moawad · S.M. Moussa · M.A. Sakr
Faculty of Computer and Information Science, Ain Shams University, Cairo, Egypt
e-mail: basma.hassan@cis.asu.edu.eg

I.F. Moawad
e-mail: ibrahim_moawad@cis.asu.edu.eg

S.M. Moussa
e-mail: sherinmoussa@cis.asu.edu.eg

M.A. Sakr
e-mail: mahmoudsakr@cis.asu.edu.eg

© Springer International Publishing Switzerland 2015
V. Popovich et al. (eds.), *Information Fusion and Geographic Information Systems (IF&GIS' 2015)*, Lecture Notes in Geoinformation and Cartography,
DOI 10.1007/978-3-319-16667-4_4

objects e.g., street names, city names, etc. There are objects that change their spatial reference with time, or so-called spatiotemporal objects. With the advancement of the current GPS technologies, large-scale capture of motion of those moving spatiotemporal objects became attainable. Typical examples of moving objects include cars and persons equipped with a GPS device, or animals wearing a transmitter whose signals are captured by satellites [2]. Understanding why and how people and animals move, which places they visit and for which purposes, what are their activities, and which resources they use, is of great importance for decision making in a variety of applications. Case in point, applications like road traffic monitoring, mobile health and animal data ecology, call for methods enabling rich and expressive representation of moving objects.

There have been works providing efficient mobile data management and mining techniques, but they focus on raw trajectories (i.e., a sequence of spatiotemporal observations (x, y, t) using geodetic coordinates). Thus, they ignore the background contextual information (e.g., transportation means and geographical objects) that can contribute in creating significant semantic knowledge about movements. Semantics refer to the contextual information available about the moving object, apart from its mere position data. Semantic is contained both in the geometric properties of the spatiotemporal stream (e.g., when the user stops/moves) as well as in the geographic space on which the object moves (e.g., shops, roads). An example of semantically enriched trajectory could be the following:

(Begin, home, 9 am) → (move, road, 9–10 am, on-bus) → (stop, office, 10 am–5 pm, work) → (move, road, 5–5:30 pm, on-metro) → (stop, market, 5:30–6 pm, shopping) → (move, road, 6–6:20 pm, walking) → (End, home, 6:20 pm)

Semantic trajectory is a growing trend that has recently emerged in geographic information science and spatiotemporal knowledge discovery. It is mainly concerned with understanding the motion of the moving object with respect to the application of interest. Adding semantics enhances the analysis of data and facilitates the discovery of semantically implicit patterns and behaviors. The community created within the FP6 GeoPKDD [3] has initiated most of the research on semantic trajectories with a special focus on privacy and security issues. Following the GeoPKDD, MODAP [4] and SEEK [5] continued the exploitation of knowledge about moving object data.

In this paper, we investigate the existing literature on semantic trajectories and propose a new classification schema for the research efforts done in semantic trajectory construction and applications till now. The proposed classification schema includes three main classes: semantic trajectory modeling, semantic trajectory computation, and semantic trajectory applications. Several similar survey efforts were presented in [6–8], but their main focus was defining the basic concepts and issues about mobility data and surveying techniques for semantic trajectory construction, annotation and knowledge extraction through mining. Our survey extends their work by covering the existing data models supporting semantic trajectory construction besides investigating the activity recognition means and modes (online and offline) for capturing spatiotemporal data. Furthermore, we present an in-depth survey of trajectory segmentation criteria and demonstrate several applications of semantic trajectories rather than the data mining. Last not least, a major contribution of this paper is the classification schema developed, which maps the existing works in the semantic trajectory research area, discussing each area separately, and identifying the challenges and the potential opportunities within them.

The rest of the paper is organized as follows: Sect. 2 presents the proposed classification schema for the semantic trajectory research work whereas Sect. 3, 4, and 5 surveys the research efforts for semantic trajectory modeling, computation, and applications respectively. Section 6 analyzes the main gaps found in the current research works. Finally, Sect. 7 concludes our conducted study.

2 Classification Schema of Studies on Semantic Trajectories

We present a comprehensive study and analysis for the current research on semantic trajectories. There are three main areas of work that exist in the relevant literature: modeling semantic trajectories, their computation, and application. The modeling area studies which part of the trajectory data will be stored, how it will be accessed, and what kind of semantics will be annotated to it. The computational area discusses the extraction of raw data, its cleaning, compression, segmentation, and annotation.

While the application area proposes different uses of the semantic trajectory data in a variety of applications. Figure 1 presents the proposed classification schema of research studies on semantic trajectories.

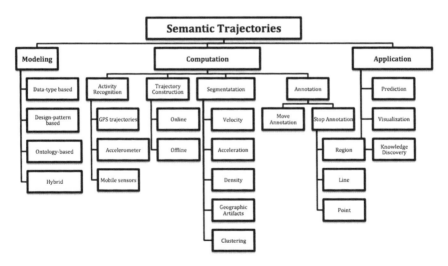

Fig. 1 A new classification schema of studies on semantic trajectories

3 Semantic Trajectories Modeling

The semantic trajectory data modeling is the main task of the semantic trajectory construction. It is the process of defining and analyzing data requirements to support the application of trajectories. There are three main levels of data models that evolve as we progress from the initial requirements to the actual database. The conceptual model maps the initial requirements as technology independent specifications. Following it is the logical data model that defines the document structures that will be used in the database. And finally, the logical data model is transformed into the physical data model, organizing data physically in the database for storage and access. In this paper we classified the semantic trajectory models, regardless of their level of abstraction, into four classes: (1) data type-based, (2) design pattern-based, (3) ontology-based and (4) hybrid data models.

3.1 Data Type-Based Modeling

The research presented in [9] introduced an algebraic model that represents a spatiotemporal trajectory (STT) as an abstract data type (ADT), encapsulating dynamic and semantic features. The ADT was designed in a way that if it got integrated in any database management system, it acquires the same status as built-in data structures. It is also supported with operations covering its spatiotemporal and semantic properties. The STT data type requires different data types varying from integer, boolean, string, enumeration, and constants to represent time, location and activities of spatiotemporal trajectories. A value of type STT is a pair $(\mathbf{A}, \mathbf{D})$ of temporally ordered sets, where $\mathbf{a}$, an element in $\mathbf{A}$ is defined as a = (l, t_s, t_e, purpose) where l $\in$ Point represents the location of the moving object, t_s and t_e $\in$ Time and purpose $\in$ Enum is the activity description. While $\mathbf{d}$, an element in $\mathbf{D}$, is a trip defined as d = (l_s, l_e, t_s, t_e, mode, path) where l_s and l_e $\in$ Point, t_s and t_e $\in$ Time, mode $\in$ *Enum* which is the movement mean and the attribute path represents the geometric semantic of the path taken. Along with the data structure proposed, they also introduced a manipulation language composed of operations on the STT data type to formulate semantic operations e.g., Activity_Before_Activity, spatial operations (e.g., STT_EndsBy_Point), temporal operations (e.g., Time_Begins_STT) and set-based operations (e.g., Union, Intersect.) A major drawback in this work is that the way the STT data type was designed made it application-dependent, as it represents the concept of space-time trajectories by a series of connected trips and activities. Yet, it provided useful data manipulation operations.

A conceptual model supporting the various requirements of the applications of semantic trajectories was still needed; a model that covers the characterization of trajectories with attributes, semantic and topological constrains and links to application objects. To fill this gap, the authors in [10] introduced *dedicated data types*.

In this research, they brought the minimal information common to all trajectories like the begin, end, moves, stops, as well as their sample points and interpolation functions, and encapsulated them in a generic data type. Whereas the application-specific information that cannot be encapsulated in the generic data type was modeled explicitly using dedicated data types. Those data types contain attributes representing the travelling object or its trajectories and have relationships linking them to the application objects.

To summarize, the ADT modeling approach is best used when the movement track is represented as a set of trips and activities. Whereas the dedicated data type modeling approach is preferable when dealing with trajectories having minimal stops, and where moves are on network-constrained paths that need basic semantics.

3.2 Design Pattern-Based Modeling

Data type modeling approaches alone are not sufficient to support the semantic trajectories application requirements. This is due to the inefficiency of using a generic data type for all application domains. In [11], the authors introduced an extensible model (i.e. trajectory design pattern) relying on the Model Analysis and Decision Support (MADS) model [12] to minimize the effort. MADS supports spatial and temporal objects and relationships (i.e. objects and relationships that have a geometry attribute describing their spatial extent and have a lifecycle attribute describing their temporal extent (the lifespan), and their activity status; active-suspended-disabled.

The trajectory design pattern aims at the explicit representation of trajectories and their components (stops, moves, begin and end) as object types in the database schema and linking those components with application objects. This model requires from the designer to add the semantic information specific to the application. The model provides the designer with a predefined sub-schema that supports the basic data structures for data modeling. Therefore, the trajectory design patterns act as a half-baked schema containing the basic objects and components of the trajectory, and show the relationships between those objects and the application objects. It is considered as half-baked, since it needs from the designer to adjust it and connect it to the rest of the application components.

3.3 Ontology-Based Modeling

Ontology is the conceptualization of a specific domain showing relationships between concepts in the form of a hierarchy. Spatial ontologies became a major research issue for most semantic-aware GIS (Geographic Information Systems) studies.

In [13], a case study was presented on the use of an ontological-based approach for modeling seal semantic trajectories. The modeling approach is based on two main components: domain ontology and time ontology. Those ontologies are a transformation of the semantic seal trajectory after developing it from the World Wide Web consortium. The ontologies represent basic domain and time concepts for the application and show the relationships between them. Along with the ontologies, rules were defined. Some are declarative (ex: travelling is an activity), and others are imperative requiring implementation using Oracle database supporting semantic technologies (ex: travelling is when maximum depth length is larger than 3 m). After that, a semantic integration between the domain and time ontologies is done using queries to understand temporal relationships.

While in [6, 14], the authors presented an ontological approach for modeling semantic trajectories, which integrated domain ontologies with spatial ontologies. It is similar to the approach mentioned earlier integrating domain ontologies with time ontologies. However, it integrates domain ontologies with spatial ontologies to answer queries based on spatial instead of temporal relationships (ex: the activity happened at which area instead of answering a query asking what activities happened during a specific time interval).

A good example of a model based on multiple ontologies is represented in [8], where the authors analyzed modeling requirements for trajectory modeling and proposed a multi-layered trajectory model. First, the raw movement data is transformed into a cleaner version called raw trajectories. These raw trajectories are then transformed into structured trajectories to get a more informative view, where segments correspond to more meaningful steps. Finally, those trajectories experience ontology mapping to add semantics. In this approach, they used three ontologies: (1) Geometric ontology, where the trajectory is perceived as the evolution of geometric location of a moving object during a given time interval, usually captured by mobile devices, (2) geographical ontology, turning the geometric polylines into something with more semantics, and (3) application domain ontology linking application domain knowledge. Figure 2 is an abstract representation we developed to illustrate the model's framework.

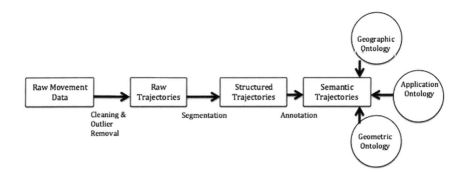

Fig. 2 Modeling using multiple ontologies

3.4 Hybrid Modeling

The proposed hybrid model in [3] encapsulated both the geometry and semantics of mobility data, supporting several levels of abstraction. It contained three models to represent the different levels of abstraction of spatiotemporal trajectories: (1) Raw data model, where raw GPS trajectories are cleaned from uncertainties and outliers to be represented as a stream of spatiotemporal tuples, (2) conceptual model, which abstracts tuples with a certain correlation (like velocity, acceleration, angle of movement, density, time interval, etc.) to become a series of non-overlapping episodes, (3) semantic model, where structured trajectories from the conceptual model are enriched with knowledge from third party sources. This research also introduced a computational platform for the progressive construction and evolution of those three models. An important contribution of this approach was that it offered a consistent framework that aimed at covering the requirements of a variety of applications, starting from those that are only interested in the raw data, to those looking for high-level of application semantic enrichments.

To summarize, choosing the right modeling approach for semantic trajectories depends on several factors. Among them is the application used, the availability of the domain's ontology, the level of trajectory abstraction required and the extent of intervention required by the database designers. *Data type-based* models are generic models that fit into a wide range of applications. They can be made persistent by extending a database model, and can be queried by extending SQL (Structured Query Language). *Design pattern-based* models are even more generic than the data type-based models, as they don't restrict to a specific data type. Instead, a dedicated type relevant to the application in hand can be added to the generic data types but will need the help of a database designer.

On the other hand, *ontology-based* models are application specific, as the ontology needs to reflect the application domain. They can represent richer semantic, and involve any kind of semantic annotations (e.g., multimedia object). In contrast to data type models, ontological models are naturally extensible because ontologies are designed to extend. Whereas the *hybrid model* is the only model that supports applications requiring several levels of abstraction, i.e., performs operations throughout the process of semantic trajectory evolution, going through the raw and structured trajectories. It also fits a wide range of applications, enabling semantic enrichment from several third party sources.

4 Computation

Semantic trajectory computation is the process of extracting and constructing spatiotemporal instances from large-scale GPS feeds, followed by semantic enrichment to comply with a predefined data. We overviewed the various stages of semantic trajectory computation, going through the activity recognition to the segmentation and annotation. Besides investigating the different modes of computation (i.e., online and offline).

4.1 Activity Recognition

This section illustrates the extraction component of the semantic trajectory computation by showcasing the activity recognition studies conducted from GPS, accelerometers, and mobile sensing device data.

4.1.1 Activity Recognition from GPS Trajectories

Many studies focus on activity recognition using GPS-based trajectory data, where the movement history of the individual is extracted in conjunction with the semantics of the location (typically from geographic or application data repositories).

To identify the important locations from GPS trajectories, studies like [15, 16] proposed methods of joining the GPS trajectory data with predefined points of interest (POI), having specific time constrains for inferring activities. For example, given a set of trajectories, a set of POIs, and an activity mapping set show possible activities that might take place and their corresponding durations; find the sequence of activities that might be performed during those set of trajectories. The rationale behind it is that if a user stays at a POI for long enough time, then some activity might take place. So, it answers questions like which POI's did the user stay in? And what activities were performed in it?

When there are no predefined POI's, a clustering method can be used as suggested by [11, 17] to automatically discover hotspots in the trajectory data. In [17], the authors discovered stops or interesting places using speed-based methods, where the distance between points is calculated along the trajectory instead of the traditional Euclidean distance. They considered the notion of minimal time instead of minimal number of points for a region to be considered dense. The *minimum time duration* indicates the minimum time necessary to generate a cluster. It is calculated by subtracting the timestamp of the first point in the cluster from the last point's timestamp in the same cluster. While in [11], the POIs were detected using the DJ-Cluster algorithm, where for each point, a neighborhood is calculated. The neighborhood consists of points within distance*Eps*, under the condition that there are at least *MinPts* of them. If no such neighborhood is found, the point is labeled noise. The DJ-Cluster algorithm has several important technical advantages: it allows clusters of arbitrary shape; ignores outliers, noise, and unusual points; has more easily chosen parameters; and has deterministic results.

The previous activity recognition studies are about the location part of trajectory data, stating, "What they move for". Another very interesting study in the literature was the recognition of the transportation modes to understand "how they move".

For example, researchers in [18] designed a methodology for detecting the transportation mode using a set of variables like acceleration, velocity, median speed, etc. Following this direction, the authors in [19] provided a more solid approach for identifying the transportation mode through a three-step framework to recognize means of transportation; first, by the segmentation of change points, second is the mode detection through a predefined decision tree and the third stage is to apply graph-based post-processing to refine the results.

4.1.2 Activity Recognition from Accelerometer Data

"A tri-axial accelerometer is a sensor that can collect a real valued estimate of acceleration along three axes, i.e., x, y and z" [6]. It has been largely used in activity recognition specifically in activities like running, walking, climbing steps, gym instruments, etc.

The most cited study in this regard was conducted in [2], where the authors were the first to use multiple accelerometer sensors worn in different parts of the body to detect common activities. The problem with this approach was that it required certain laboratory conditions, i.e. not easily applicable in normal circumstances. Further research has been developed in [20], making this approach more user friendly and enhancing its mobility by only using one accelerometer.

4.1.3 Activity Recognition from Mobile Phone Sensors

Mobile phone sensors activity recognition is done through the use of wireless devices like smart phones to understand what people do, where they go, and how they interact with each other. Combining accelerometer data with mobile phone audio data through a microphone to better detect the activity is an example. Several studies have been conducted in this field [1, 21, 22], where they used smart phone embedded sensors and data records, like GPS, GSM cell tower, call and SMS logs, Wi-Fi, Bluetooth, accelerometer, and audio features for mining people's activities.

4.2 Trajectory Construction

The semantic trajectory construction is the process of integrating the spatiotemporal movement characteristics with useful information regarding objects movement patterns and social activities. There are two modes for semantic trajectory construction: (a) Offline mode, where all trajectory construction processes are done offline, and (b) online mode, where parts of the trajectory construction processes are done in real time.

4.2.1 Online Mode

In the current literature, timely trajectory computation to serve real-time queries for today's trajectory applications (ex: traffic monitoring) is not sufficient. To fill this gap, the authors in [23] proposed SeTraStream, a platform for online semantic trajectory construction. The main contributions of SeTraStream can be summarized as follows:

1. Online trajectory preprocessing: trajectory preprocessing was redesigned to include online cleaning using the kernel smoothing method, and online compression using the synchronized Euclidian distance and correlation coefficient.
2. Online trajectory Construction, where they designed techniques for episode identification during the online trajectory segmentation. Some of the above offline works can adapt to an online context. Yet, none of them support the exploitation of the profound semantics that exist in the computed trajectories in real-time.
3. Platform implementation and evaluation: an online framework that enables semantic trajectory construction over streaming movement data tackling real time streaming environments.

The flow is as follows: The server receives from the mobile object device a batch of GPS data with a predefined window size and stream complementary features, like acceleration, speed, displacement, etc. Consequently, cleaning, smoothing and compression techniques are applied. Finally, feature vectors are extracted, and a corresponding matrix is formed and the batch is buffered until segmentation takes place. During the segmentation, previously buffered batches are de-queued and matched with dissimilar batches (based on RV-coefficient) to form an episode. Having detected an episode, SeTraStream defines the triplet (semantic tagging) describing its start and end time bounded to a specific geometry. With this mode of computation, new challenges to the conventional methods came to existence. As in the offline mode algorithms, threshold tuning is common. While in the online context, parameter tuning is prohibitive. In Fig. 3, we represented the flow of the online trajectory construction.

4.2.2 Offline Mode

In this mode, the movement data in the form of large-scale GPS datasets is collected in advance. The processing undergoes several stages starting from data refinement, tuning, map matching and compression to trajectory identification, and eventually trajectory segmentation and annotation. The offline trajectory-computing framework used for a specific application should reflect its semantic trajectory modeling requirements. For example, the authors in [3] designed an offline trajectory computational framework matching the requirements of the hybrid spatiotemporal model they proposed. The framework is composed of three layers:

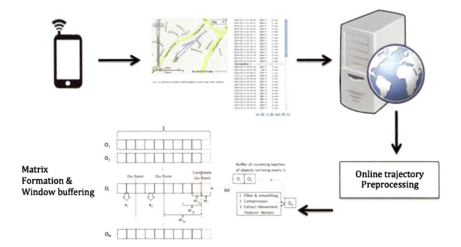

Fig. 3 Flow of online trajectory construction

1. The data preprocessing layer, where the outlier removal, kernel smoothing and compression stages occur. Several works have been conducted in this specific area as in [10, 24–27].
2. The trajectory identification layer, which is responsible for dividing the processed GPS raw data into trajectories using different policies (ex: GPS gap, predefined time interval, predefined space extent) [3, 8 28].
3. The trajectory structure layer that works on the identified trajectories. It further divides them into episodes i.e., meaningful stop and moves ready for semantic tagging/annotation using geographic artifacts, speed, velocity and direction based methods [3, 15, 29, 30].

In [8], a similar computational model was used, adding the semantic enrichment stage to the trajectory structure layer. It was customized for the multi-layered model mentioned earlier by linking the spatiotemporal units with semantic knowledge from the geographic data and application domain data.

Research is still needed to substantially reduce the amount of raw data, while not missing valuable information. On the fly analysis techniques are also required for data processing. This is because it is unaffordable to store first then reduce afterwards with the data's exponential inflation nature. Existing work also assumes well-recognized constraints on valid data or well-understood error models; but for many emerging big data domains, these do not exist.

4.3 Segmentation

The authors in [31] proposed the first data model looking at the trajectories from a conceptual point of view, where they divided the trajectory into a set of stops and moves. From this starting point, different works have been proposed to instantiate the model of stops [30, 32]. A stop can be defined as "the important places where a trajectory has passed and stayed for a reasonable time duration" [6]. For this kind of segmentation, different approaches have been proposed as follows.

4.3.1 Velocity-Based

The velocity-based approach [6] focused on stops and moves, where it determines if a GPS point belongs to a stop or to a move episode by using a speed threshold. Hence, if the instant speed of p is lower than the threshold, it is a part of a stop; otherwise it belongs to a move.

4.3.2 Density-Based

Using only velocity for identifying stops is not enough for some scenarios. Therefore, the authors in [6] designed a density-based stop discovery approach. It considered not only the speed but also the maximum diameter that the moving object has traveled during a given time duration.

4.3.3 Geographic Artifacts

Trajectories and geographic data overlap in space. In [15], the authors integrated geographic data with sub-trajectories overlapping in geometry. This is done in a user-dependent way, where the user identifies which places are of interest to his specific application to disregard any geographic places out of the application's interest. They devised the algorithm SMoT (Stop and Moves of Trajectories) that verifies for each point of the trajectory if it intersects the geometry of a candidate stop (i.e. a geographical place related to the application) and that the duration of the intersection is at least equal to a specific predefined threshold.

4.3.4 Clustering based

An extension to SMoT [15] was developed in [30] using the method CB-SMoT, which stands for Clustering Based—SMoT. It used a clustering technique in order to identify stops according to Spaccapietra's stop definition. In [22], instead of comparing each and every point with the geometry of the geographic place, clusters of trajectories were identified beforehand according to their speeds and then they were mapped to geographic places to add semantics to those clusters.

4.4 Annotation Approach

This is the stage where trajectories are transformed into semantic trajectories in the computation stage. It is the task following the trajectory segmentation where meaningful information is assigned to specific intervals and sections of the moving object's movement track.

4.4.1 Annotating Moves

The annotation techniques mentioned above were mainly concerned with annotating the stops defined in [31] or annotating trajectory episodes introduced earlier in [33]. They defined an episode as "a discreet time period for which the user's spatiotemporal behavior was relatively homogeneous". Very few research works [5, 23] had their focus mainly on annotating moves. Annotating moves is necessary because not every stop in the physical trajectory possesses (application dependent) interpretation. The semantic stops can happen without appearing in the data.

4.4.2 Stop Annotations

Stopping in a trip means that there is something of interest to do. So stop annotation is about mapping stops to places of interest, which can be geographical regions, roads in the form of lines, or POIs in the form of points.

a. **Regions**: Annotating trajectories with regions of interest from geographical or application domain sources. It does so by computing topological correlations between trajectories and 3rd party data sources containing semantic places of regions [3, 6, 15].
b. **Lines**: It is the annotation of trajectories with lines of interest like road networks. Given data sources of different forms of road networks, the purpose is to identify correct road segments, as well as, infer transportation modes such as walking, cycling, and public transportation like metro e.g., [19, 34].
c. **Points**: It is the annotation of stop episodes of a trajectory with information about a suitable point of interest. Examples are shown in [4, 5, 32]. However, densely populated urban areas bring several candidate POIs for a stop. In addition, low GPS sampling rate (due to battery outage and GPS signal losses) makes the problem more intricate. Therefore, the authors in proposed the Hidden Markov Model (HMM)-based technique for semantic annotation of stops, which was able to overcome those problems. In the Hidden Markov Model, the state is not directly visible, but the output, dependent on the state, is visible. Each state has a probability distribution over the possible output tokens. Thus, the sequence of tokens generated by an HMM gives some information about the sequence of states. It can be presented as the simplest dynamic bayesian network.

5 Semantic Trajectory Applications

Adding meaning to the movement track of moving objects opened new perspectives for a large number of applications built on the semantic of movements of objects. This section classifies the state of the art applications into trajectory prediction, visualization and knowledge discovery applications.

5.1 Prediction

Many applications, such as location-based advertisement, navigational planning services and traffic management, have been developed for the location-based services market. Those applications require accurately predicting the next move of the moving object. The first to predict destinations from partial trajectories where the authors in [35] described a method called predestination that uses the history of a driver's destinations to predict how his trip will progress later on. Another example of prediction was a model developed in [36] where prediction was based on social spatial approximation, which utilizes current GPS coordinates of user friends to estimate GPS coordinate of the user. The authors in [37] proposed a novel approach named GTS-LP for mining and prediction of mobile user's movement behavior. They defined a new pattern, called the GTS-Pattern, to represent frequent moves, which based on it they proposed the location prediction strategy.

5.2 Visualization

An effective way for semantic trajectories analysis is to visualize the movement track. In [38], the main plot area used to visualize the trajectories was a 3D cube with three axes, the x-y geographical location and the time axis, where trajectory data and domain ontology were mapped into 3D cubes. Another research was conducted using Weka-STPM [39], with new pre-processing methods and a graphical GUI to visualize in a map the spatial entities and the generated stops and moves. Another example of a system enabling trajectory visualization is MoveMine [40], which provides a user friendly interface where users can select a data set and the corresponding raw data is plotted on the Google Map. Furthermore, a user can plot the results in Google Earth for 3-D visualization of the results.

5.3 Knowledge Discovery

There are approaches that exploit semantic trajectories for knowledge discovery, in particular movement patterns. Among them is [41], which proposed a novel methodology for recognizing the behavior of moving objects within stops. This was done by further dividing a stop into sub-stops using velocity/direction based rules.

While in [34], the authors developed a pattern mining framework which detected moving patterns between two stops considering background geographical information, e.g., pattern of movement of tourists between touristic places. Several other works [15, 30] developed similar pattern and knowledge mining techniques for a pool of applications, ranging from identifying tourists' POIs to understanding moving object behaviors and trajectory goals.

Furthermore, a scalable reference framework for the semantic management of moving objects called SemanticMOVE was proposed in [42], which supports better mining, analysis and reasoning of semantic mobility data. It's a generic architecture with an infrastructure of distributed nature where each object collects, stores, processes and analyzes the semantics of its own data.

6 Research Gaps

During our extensive study of semantic trajectories, several research gaps have been deduced throughout the previous studies and literature concerned with semantic trajectory construction and application. These gaps include:

- The data type-based models need to be less application-dependent and more generic to include the wide range of scientific domains, besides the advancement in manipulation languages for querying and knowledge discovery.
- In ontology-based modeling, research on applying more domain conditions on rules is becoming a necessity to reduce time and space storage inference complexity.
- In semantic trajectory extraction and activity recognition, more research is needed to address their use, and how they can be integrated with online computational platforms and geographical maps.
- There is a huge research opportunity in the area of trajectory segmentation using means rather than the episodes and stop and moves identification models.
- More research is required to focus on annotating moves, because a huge part of the semantics of moving objects lies in the movement activity rather than activities done at stops, besides adding to the logic behind the semantics at stops. Also, better stop analyses can be made via careful tuning (e.g., tuning stop identification and interpretation to make it work even for short stops).
- To the best of our knowledge, online mode algorithms for semantic trajectory construction are significantly missing. In current online mode research, the tagging needs to be customized according to different application contexts by modifying the feature vector (with features like segment distance, duration … etc.), besides using the corresponding suitable tagging technique including decision trees, neural, and bayesian methods.
- We are living in the era of 'Big Data'. Spatiotemporal trajectories, whether captured through remote sensors or large-scale simulations, has always been

'Big'. However, recent advances in instrumentation and computation made spatiotemporal data even bigger, putting several constraints on data analytics capabilities. Spatial computation needs to be transformed to meet the challenges posed by the big spatiotemporal trajectories.
- For semantic trajectory application, more innovative research is also expected through integrating traditional knowledge extraction techniques with visualization approaches, and with knowledge extracted from social network interactions.

7 Conclusion

In this paper, we discussed the main components of the semantic trajectory processing by analyzing the state of the art and past research contributions in this field. The relative novelty of the domain leaves many challenges, opportunities and extended studies open for future work, which we addressed most of them in our deduced research gaps. We were able to conclude the analysis and insights of our study as follows: (1) Starting with the trajectory extraction component, most of the literature focused on the conventional GPS tracking devices disregarding the wide penetration of smart phones that can be used for a broader range of applications in real time context, (2) from a data modeling perspective, several spatiotemporal models have been developed to include the semantic dimension. The hybrid models are the only variant that support different levels of data abstraction by representing trajectories in terms of both spatial and semantic mobility characteristics, (3) an essential component of semantic trajectory construction is the segmentation. The most common method of segmentation is the stop and move, which was the basis of many studies focusing on stop discovery techniques relying on speed, velocity, acceleration, direction, geographic artifacts and clustering algorithms, (4) research in semantic annotation of trajectories is either in annotating moves or in stop annotation, where stops are characterized as regions, lines, or points, (5) semantic trajectory applications fall in three main categories; knowledge discovery, visualization and prediction. There is a need to develop applications targeting large and deforming objects (e.g., oil spills, diseases ... etc.), network-constrained movements, relative movement, and collective movement for any kind of collections of objects, and finally (6) we have given an extensive survey of works done on the aspects of semantic trajectories. We have also highlighted research gaps in those areas to call for future work.

References

1. Yan Z, Chakraborty D, Misra A, Jeung H, Aberer K (2012) Semantic activity classification using locomotive signatures from mobile phones
2. Bao L, Intille S (2004) Pervasive computing, vol 3001. Springer, Berlin, pp 1–17
3. Yan Z, Parent C (2010) The semantic web: research and applications, vol 6088. Springer, Berlin, pp 60–75

4. Yan Z, Chakraborty D, Parent C, Spaccapietra S, Aberer K (2011) SeMiTri. In: Proceedings of the 14th international conference on extending database technology—EDBT/ICDT '11, p 259
5. Yu F, Ip HHS (2008) Semantic content analysis and annotation of histological images. Comput Biol Med 38(6):635–649
6. Zhixian Y (2011) Semantic trajectories: computing and understanding mobility data In: Ph.D. dissertation, Swiss Federal Institute of Technology, Information and Communication Dept., Lausanne
7. Parent C, Pelekis N, Theodoridis Y, Yan Z, Spaccapietra S, Renso C, Andrienko G, Andrienko N, Bogorny V, Damiani ML, Gkoulalas-Divanis A, Macedo J (2013) Semantic trajectories modeling and analysis. ACM Comput Surv 45:1–32
8. Yan Z, Spaccapietra S (2009) Towards semantic trajectory data analysis: a conceptual and computational approach. VLDB PhD Work 15(2):165–190
9. Zheni D, Frihida A (2009) A semantic approach for the modeling of trajectories in space and time. In: Advances in conceptual modelling, pp 347–356
10. Marketos ARG, Frentzos E, Ntoutsi I, Pelekis N, Raffaetà A, Theodoridis Y (2008) Building real-world trajectory warehouses. In: MobiDE
11. Zhou C, Frankowski D, Ludford P, Shekhar S, Terveen L (2004) Discovering personal gazetteers. In: Proceedings of the 12th annual ACM international workshop on geographic information systems—GIS '04, p 266
12. Parent C, Spaccapietra S, Zimányi E (2006) Conceptual modeling for traditional and spatio-temporal applications: the MADS approach. Springer, New York, p 450
13. Wannous R, Malki J, Bouju A (2013) Time integration in semantic trajectories using an ontological modelling approach A case study with experiments, optimization and evaluation of an integration approach. Springer, Berlin
14. Wannous R, Malki J, Bouju A, Vincent C (2013) Modeling approaches and algorithms for advanced computer applications, vol 488. Springer International Publishing, Cham
15. Alvares LO, Bogorny V, Kuijpers B, de Macedo JAF, Moelans B, Vaisman A (2007) A model for enriching trajectories with semantic geographical information. In: Proceedings of the 15th annual ACM international symposium on advances in geographic information systems—GIS '07, p 1
16. Xie K, Deng K, Zhou X (2009) From trajectories to activities. In: Proceedings of the 2009 international workshop on location based social networks—LBSN '09, p 25
17. Zhao XL, Xu WX (2009) A clustering-based approach for discovering interesting places in a single trajectory. In: 2009 2nd international conference on intelligent computing technology and automation, ICICTA 2009, vol 3, pp 429–432
18. Schuessler N, Axhausen KW (2009) Processing raw data from global positioning systems without additional information. Transp Res Rec J Transp Res Board 2105(1):28–36
19. Zheng Y, Chen Y, Li Q, Xie X, Ma W-Y (2010) Understanding transportation modes based on GPS data for web applications. ACM Trans Web 4(1):1–36
20. Ravi N, Dandekar N, Mysore P, Littman M (2009) Distributed computing, artificial intelligence, bioinformatics, soft computing, and ambient assisted living, vol 5518. Springer, Berlin
21. Kiukkonen N, Blom J, Dousse O (2010) Towards rich mobile phone datasets: Lausanne data collection campaign. Proc ICPS, Berlin
22. Choujaa D, Dulay N (2010) Predicting human behaviour from selected mobile phone data points. In: Proceedings of the 12th ACM international conference on Ubiquitous computing—Ubicomp '10, p 105
23. Yan Z, Giatrakos N, Katsikaros V (2011) Advances in spatial and temporal databases, vol 6849. Springer, Berlin, pp 367–385
24. Schüssler N, Axhausen KW (2009) Processing GPS raw data without information. Transp Res Rec J Transp Res Board 8:28–36
25. Douglas D, Peucker T (1973) Algorithms for the reduction of the number of points required to represent a digitized line or its caricature. Can Cartogr 10(2):112–122

26. Meratnia N, de By RA (2004) Spatiotemporal compression techniques for moving point objects. In: EDBT
27. Potamias M, Patroumpas K, Sellis T (2006) Sampling trajectory streams with spatiotemporal criteria. In: 18th international conference on scientific and statistical database management (SSDBM'06), pp 275–284
28. Yan Z, Chakraborty D, Parent C, Spaccapietra S, Aberer K (2013) Semantic trajectories. ACM Trans Intell Syst Technol 4(3):1
29. Yan Z, Spremic L, Chakraborty D, Parent C, Spaccapietra S, Aberer K (2010) Automatic construction and multi-level visualization of semantic trajectories. In: Proceedings of the 18th SIGSPATIAL international conference on advances in geographic information systems—GIS '10, vol 12, p 524
30. Xiu-Li Z, Wei-Xiang X (2009) A clustering-based approach for discovering interesting places in a single trajectory. In: 2009 second international conference on intelligent computation technology and automation, pp 429–432
31. Spaccapietra S, Parent C, Damiani ML, de Macedo JA, Porto F, Vangenot C (2008) A conceptual view on trajectories. Data Knowl Eng 65(1):126–146
32. Vald F, Damiani ML, Güting RH, (2010) The semantic web: research and applications, vol 6088, Part 2. Springer, Berlin, pp 450–453
33. Mountain D, Raper J (2001) Modelling human spatio-temporal behaviour: a challenge for location-based services. In: Proceedings of the sixth international conference on geocomputation
34. Alvares LO, Bogorny V, de Macedo JAF, Moelans B, Spaccapietra S (2007) Dynamic modeling of trajectory patterns using data mining and reverse engineering. In: Tutorials, posters, panels and industrial contributions at the 26th international conference on conceptual modeling, vol 83, pp 149–154
35. Krumm J, Horvitz E (2006) Predestination: inferring destinations from partial trajectories. In: Proceedings of the 8th international conference on Ubiquitous computing (UbiComp '06), pp 243–260
36. Backstrom L, Sun E, Marlow C (2010) Find me if you can: improving geographical prediction with social and spatial proximity. In: Proceedings of the 19th international conference on world wide web, pp 61–70
37. Ying JJ-C, Lee W-C, Tseng VS (2013) Mining geographic-temporal—semantic patterns in trajectories for location prediction. ACM Trans Intell Syst Technol 5(1):1–33
38. Bakshev S, de Macêdo J, Spisanti L (2010) Semantic visualization of trajectories. In: ICEIS, pp 1–10
39. Alvares L, Palma A, Oliveira G, Bogorny V (2010) Weka-STPM: from trajectory samples to semantic trajectories. In: Proceedings of the Workshop Open Source Code, vol 1
40. Li Z, Ji M, Lee J, Tang L, Yu Y, Han J, Kays R (2010) MoveMine. In: Proceedings of the 2010 international conference on management of data—SIGMOD '10, p 1203
41. Moreno B, Times V, Renso C, Bogorny V (2010) Looking inside the stops of trajectories of moving objects. In: Geoinfo
42. Ilarri S, Stojanovic D, Ray C (2015) Semantic management of moving objects: a vision towards smart mobility. Expert Syst Appl 42(3):1418–1435

Semantic Recommender System for Touristic Context Based on Linked Data

Luis Cabrera Rivera, Luis M. Vilches-Blázquez, Miguel Torres-Ruiz and Marco Antonio Moreno Ibarra

Abstract The lack of personalization presented in touristic itineraries that are offered by travel agencies involve a little flexibility. Basically, they are designed with the points of interest (POIs) that have more relevance in the area. On the other hand, there are POIs that have agreements with the agencies, which originate a excluding POIs that could be interesting for the tourist. In this work, a method capable to use the user preferences, like POIs and activities that user wants to realize during their vacations is proposed. Moreover, some weighted features such as the max distance that user wants to walk between POIs, and opinions of other users, coming from the web 2.0 by means of social media are taken into account. As result, a personalized route, which is composed of recommended POIs for the user and satisfied the user profile is provided.

1 Introduction

Nowadays, there are several travel agencies around the world that design tours according to the most relevance Point of Interests (POIs), in a certain order and with estimated time. The itineraries do not take into consideration the available time of users for visiting, because the tour is pre-designed with the best-known POIs such

L. Cabrera Rivera (✉) · M. Torres-Ruiz · M.A. Moreno Ibarra
Centro de Investigación En Computación, Instituto Politécnico Nacional UPALM-Zacatenco, 07320 Mexico, Mexico
e-mail: gir250@gmail.com

M. Torres-Ruiz
e-mail: miguel.torres.ruiz@gmail.com

M.A. Moreno Ibarra
e-mail: mmorenoi@ipn.mx

L.M. Vilches-Blázquez
National University of Colombia, Carrera 30 No 45A-03, Bogota D.C, Colombia
e-mail: lmvilches.blazquez@gmail.com

© Springer International Publishing Switzerland 2015
V. Popovich et al. (eds.), *Information Fusion and Geographic Information Systems (IF&GIS' 2015)*, Lecture Notes in Geoinformation and Cartography,
DOI 10.1007/978-3-319-16667-4_5

as museums, restaurants and archaeological places, among others. However, there are users who want to know other interesting POIs, and not just a specific place. For example, a person wants to know some museums in the city, and he is only interested in "baroque art museums", or another person wants to have dinner in a restaurant, but he is only interested in "Japanese food", and he wants the closest one. After that, he wants to visit the most popular dancing place.

In order to take into account the user preferences, a user profile should be used, which is a representation of essential information about the user [1], where the compiled information depends directly on the application domain that will be used. In this case, the retrieved POIs and activities are the particular characteristics that user wants to consider according to their interest.

Regarding the requirements for personalization, there are an increasing number of systems that can suggest things to users for e-commerce, web pages, or recommended contents based on their profile; this kind of systems are called *recommender systems* [2]. These systems allow us to make an analysis based on user preferences in order to provide suggestions. Thus, they design filters for improving the accuracy in the recommendations. There are different types of filters that are implemented in the systems and granted best results depending on each one. In the case of taking into account other user opinions a collaborative filters are defined [3]. These filters allow systems use evaluations performed for other users and so be able to generate a recommendation with more accuracy for the user about some objects according the preferences of similar users.

Recommender systems are related to Location-Based Services (LBS), which can help to set the position of a user and then be used with the external information for providing personalized application and services [4]. In the tourism context, the location-based services aid to retrieve the current position of a user as well as the position of recommended POIs. Up-to-date, there are several applications that provide location-based services as well as brief descriptions of the POIs, such as SMARTMUSEUM [5], SPETA [6], SigTur/E-Destination [7], DBpedia Mobile [8], and applications focused on social media like Foursquare [9], which allows seeing users who visit a POI, with opinions that may be good or bad about the service or aspect of the visited POI. In addition, this social media allows us to add new POIs or search a specific kind of POI.

There are also works like [10, 7], in which the use of semantic analysis for giving a formal and semantic description about a certain object are proposed. That description is recommend to the users and the analysis disambiguates recommendations to a single recommendation that can proceed from different recommendation systems.

For that purpose, an ontology defined as an explicit specification of a conceptualization [11] is proposed. The ontology development focused on modeling the user profile is described in [12, 13, 7]. An ontology provides an enriched description about objects that are recommended [14], because the ontology contains properties to the object, like "name", "price", "category", etc. They give to the description a meaning in the real world.

In this work, a recommender system that exploits semantic information based on to Linked Data principles is proposed. This ongoing application consists of a semantic approach for recommending POIs in a tourism context. This approach considers different opinions retrieved from users within several social media (like Foursquare), semantics aspects (like the concept to which it belongs) and geographic aspects. We also include a user profile in order to retrieve the user preferences and process them for retrieving a list of activities and POIs that represent everything that user wants to do.

As case study the first square of the historic Center of Mexico City was taken. This area offers a great diversity of POIs such as restaurants, malls, bars, museums, pictorial files, and historical monuments. These POIs were considered by their scores, which were obtained from the social media. Thus, each POI has been described into ontology.

This paper is structured as follows. Section 2 presents the state of the art related to the work in this field. Section 3 presents the proposed approach to recommend POIs and semantically generate tourist itinerary. Section 4 depicts the experimental results obtained by applying our approach. Finally the conclusion and future works are outlined in Sect. 5.

2 Related Work

Recommender systems have been developed to make suggestions about certain objects of interest to the user, for example: movies, magazines, video games, web pages, among others. It retrieves information about the user preferences as a set of objects (books, types of food, applications, etc.) [2]. This information can be retrieved explicitly when the user fills a form with preferences, and implicitly when the system has to mining information about the user from social media, as visiting web pages, listening to music, etc. [12, 15].

This kind of systems usually are employed in e-commerce [16], because the systems provides a great accuracy in their recommendations about things that may be of interest to the user among millions of objects, which are offered in a web page (i.e. Amazon [17]). Recommender systems employ filter algorithms for analyzing the user preferences. According to [18–20], the filters are classified as follows: (1) *Content-based,* which uses the recommendations based on previously decisions taken by the user. It uses a similarity measure to determine which other objects can be recommended to the user. (2) *Demographic-based* uses certain characteristics such as gender, age or country to determine preferences in common with other users, who share similar characteristics. (3) *Collaboration-based* uses the opinions of users to a set of objects (movies, attractions, video games, etc.), for storing and collecting the amount of information needed to generate better recommendations for the users, which are based on the assessments by other users with similar preferences.

On the other hand, geographic ontology can be defined as a double perspective [21], first like a general and technical vision that emerges from the scope of ISO/TC 211, and second with a more specific perspective linked to the geospatial domain. ISO/TC 211 asserts that geographic domain ontology refers to the formal representation of a phenomenon with an underlying vocabulary. This representation includes definitions and axioms that make explicit the intended meaning and used to describe the phenomenon, and those relationships that can be used by software applications to support sharing, reuse and integration of geographic information with any other source of information within knowledge of a domain and across domains of knowledge.

From a perspective related to the domain, in [22] the concept of geo-ontology has two basic types of definition. The first one corresponding to physical phenomena in the real world and the second one corresponds to real world phenomena that have been created to represent institutional and social structures. Thus, the concepts in a geo-ontology are directed towards spatial objects in the world. In [23], a study on relevant ontologies for using in qualitative reasoning and interoperability in geospatial intelligence community is presented.

On the other hand, touristic ontologies are classified within geospatial ontologies, which are composed of geographic objects in tourism context. These ontologies are ideal for an interpretation of sites and activities that can be performed in a specific area. Currently, there are several ontologies in this domain, which consists of global standards and thesaurus of the World Tourism Organization and Gazetteers that serve as vocabularies for building these ontologies that are used in recommender systems. Examples of these ontologies are Harmonise Ontology [24], Mondeca Tourism Ontology,[1] Hi-Touch (see footnote 1), and OnTour Ontology.[2]

SMARTMUSEUM [5] is an application implemented on mobile devices and web. This system suggests cultural and artistic places, and gives a description of the landmark as well as videos and images. The system uses a content filter and languages of Semantic Web for performing ubiquitous computing to represent data. An ontology for heterogeneous content descriptions of POIs is used to retrieve position using GPS for exterior and RFID for interior. It considers using a user profile for preferences, and a framework for retrieving information on the contextual data.

In SPETA [6] a hybrid filter is implemented. A touristic ontology for giving a semantic context to the recommendations is defined. It also uses some social media to retrieve implicit information of user profiles for their analysis, specifically to visit places and preferences, including interactions with other users. The hybrid filter employed consists of three stages: (1) Take the contextual data like weather, time and current location of the user in order to determine all the services, which can be offered. (2) It uses a knowledge base to use the tourist ontology; it will model the user preferences and services that could be offered. This technique employs a

---

[1] http://www.mondeca.com.
[2] http://e-tourism.deri.at/ont/index.html.

similarity measure based on features for retrieving user preferences and services. (3) It retrieves preferences from social media to know the things that user and his friends prefer and take into consideration for the recommendation.

SigTur/E-Destination system [7] employs for its recommendations a collaborative filter and similarity measures that collect assessments from other users. This coupled with various methods that help to determine which activities the user considers. DBpedia Mobile [8] is composed of a DBpedia client of centric locating for mobile devices, which consists of a base map and Linked Data browser based on Fresnel. The system allows the user access to information about DBpedia resources in its proximity, from which they can access to other resources on the Semantic Web by means of links. The map that shows the information is built by RDF triples, obtained from the current area, language and weights for filtering to the server.

3 An Overview of the Semantic Recommender System

In this section, the process for generating a semantic touristic route is described. In Fig. 1, we present the core of our semantic recommender system.

According to Fig. 1, the first block is focused on the analysis and processing of the original data. In this block, we transform the original data into a RDF file. The second block is the user profile characterization that is in charge of analyzing the request form, making for the client and decomposing into three sections. The first one the activities that the client wants to make, the second one the position where is the client and the third is the weighted parameter (distance and ranking). These parameters are used to produce a weighted vector, which has all the variables and it is sent to the server for the implemented spatial operations.

The third block is focused on the geoprocessing analysis and routing service. In this block, the weighted vector is taken and a spatial operation is performed,

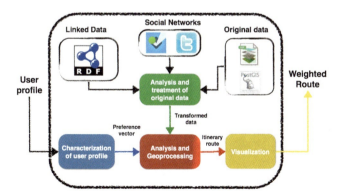

Fig. 1 General framework of the methodology

depending on the request made by the user. The spatial operations that were implemented are the follows: buffer, intersection and containment. The result of the spatial operations is a set of the recommended POIs with theirs geographic component. The last block is in charged of visualizing the selected POIs that will be displayed into the map for user visualization.

There are some proposals focused on using Bayesian networks to model the possible activities [25]. It is based on characteristics such as age, occupation and personality; according to the behavior in order to determine the type of travelers, their motivation for computing the prefer activities.

On the other hand, this work uses the analytic hierarchy process to establish the ranking of the tourist attractions. This approach consists of four steps: (1) Build a decision matrix with the values of each criterion; (2) The construction of a matrix of pairwise comparison of criteria; (3) Obtain the relative weight of the criteria of the pairwise comparison matrix; and (4) Compute the range of each alternative based on the relative weight derived. Finally, information of attraction along with the time spent at each attraction; besides using an external spatial web service for computing routes between attractions is displayed in a web page.

3.1 Data and Modeling

The assumption in this work consists of the information sources that are located from different repositories and diverse formats. Taking into account this scenario, data require a previous transformation of them in order to harmonize them.

The first information source comes from a spatial database that stores all geographic object contained in the shapefiles, which contains the streets of the area, represented by *linestrings*, where each line has a street name, an ID, its geometric and attributive components. These files also contain POIs of the area, and they were retrieved from the Mexican government web page. POIs are represented by points and also have a description, address, schedule and its geometric component.

The second information source comes from the web 2.0 (Foursquare) and web 3.0. Thus, we collected from web 2.0 the ranking and ID assigned to each POI registered in Foursquare for later ranking updates.

For modeling aforementioned data, we have developed an ontology network. According to [26], an ontology network is defined by the collection of ontologies related among a variety of different relations such as mapping, modularization and versioning. In our case, the ontology network is described in RDF format in order to be stored in a *triple-store*. The ontologies only contain concepts, relations and type of data without instances. In Fig. 2 the ontologies that compose the ontology network are depicted.

According to Fig. 2, the ontology network has been developed using the NeOn methodology [27] and is composed by three modules. The tourism module is based on the thesaurus created from the World Tourism Organization with the cooperation of the French government. This ontology was adapted according to the existing

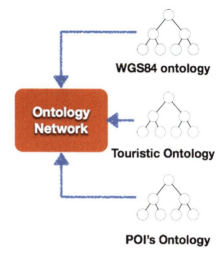

Fig. 2 The ontology network of this study case

concepts for the case study. So, the ontology has been designed according to the METHONTOLOGY methodology. On the other hand, WGS84 is a basic RDF vocabulary that provides the Semantic Web community with a namespace for representing lat(itude), long(itude) and other information about spatially-located things, using WGS84 as a reference datum. Finally, the POIs module collects multiple points of interest, as well as activities and events associated with touristic activities. This ontology network is stored in a SPARQL Endpoint, and this endpoint will be populated by the transformation process of the original data of the first source to RDF file.

3.2 Generating Tourism Linked Data

The first process to generate Linked Data information is to transform the first information source to RDF files, and then when we have both the street map and points of interest, a transformation process of these files to RDF format is performed. The shapefile information is stored in a spatial database using the QGIS framework, once it is stored; a plug-in was developed in Java. This plug-in reads the table tuples of the spatial database through spatial queries, the retrieved information from queries are name, ID and some existing description. In the case of POIs, the latitude and longitude are extracted in conjunction with its name, address, schedule, ID and concept that belongs for each POI. In the case of the streets, the initial and ending point of the segment, the street name and the ID are extracted.

Later, a URI for each instance is defined, thus, the retrieved fields in the query are coupled to form each entry in the resultant RDF file. Finally, there are two RDF files, one for the street map and other for POIs. Both files will be stored in a

SPARQL Endpoint that was implemented in a Parliament[3] triple store. For the POIs RDF file, a union with the structure of the tourism ontology is produced. Next, we can see an instance of POI that has a concept URI:

```
<rdf:Description
rdf:about="http://datos.itinerariosTuristicos.mx/recurso/Hotel_Azores">
  <rdf:type
rdf:resource="http://datos.itinerariosTuristicos.mx/ontologia/Hotel"/>
    <geo:hasGeometry
rdf:resource="http://datos.itinerariosTuristicos.mx/recurso/geometry/
Id33"/>
  <rdfs:label xml:lang="es">Hotel Azores</rdfs:label>
</rdf:Description>
```

In the last code, the URL "http://datos.itinerariosTuristicos.mx/ontologia/Hotel" is proposed to act like unique URI that links the "Hotel Azores" to its concept. So, a population task can be made in a semi-automatic way. The definition of its geometry is performed too.

3.3 User Profile Characterization

To process the user preferences (POIs, activities, distance and other users opinions), is needed a feature extraction process. Such process begins when the user performs a certain request from the application; this generates a text with XML format, which encapsulates the desired POIs and activities to be visited. By taking into account the user position and the weighted features (distances and ranking) the XML text is used to generate and send a SOAP request to the server. Once the server received the request, it proceeds to extract the preferences and performs a vectorization according to the vector prototype called preference vector. The preferences vector is formed with the parameters required to filter information (see Fig. 3).

3.4 Exploiting and Displaying Geospatial Characteristics

Once the preferences vector is generated, we proceed to extract from the SOAP request which function will be performed in the server. In our case, we will develop six possible processes that will be described in detail in the next sections, they are divided into "Simple searches" and "Spatial operations".

---

[3]http://parliament.semwebcentral.org/.

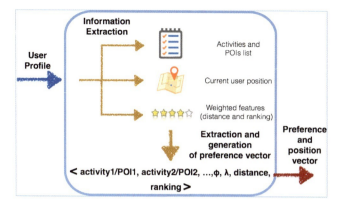

Fig. 3 General process to generate the preference vector

3.4.1 Simple Searches

In this kind of searches, we use two spatial queries for performing each process when the queries are performed. So, the map displays the POI with its corresponding landmark and a detail button. The button performs another query, which extracts from other repositories detail information and place images. The types of queries that are developed in our methodology are described as follows.

- *Search by name.* This search retrieves all labels (names) that exist in our ontology network. The second query is performed, considering the selection. The obtained result is the name of the POI, its geographic position, and its category.
- *Search by categories.* It retrieves all the categories defined as concepts in our ontology network. The second query is based on the selection, and the obtained result is all the POIs that belong to the category, its geographic position and its category are retrieved too.
- *Search by activity.* The first one is for retrieving all the activities defined in our ontology network. The second query is based on the selection, and the result is all the POIs where the user can perform the selected activity, its geographic position and category.

3.4.2 Spatial Operations

Operations performed in the server use defined functions that were implemented, and some spatial operations for retrieving POIs are developed too.

- *Search of POIs by distance ratio.* For this function, we implement a buffer operation to retrieve the POIs inside the defined ratio; the required parameter is the geographic position. Now with all data the server proceeds to execute a

query to SPARQL Endpoint where the ontology network is stored, as result the server retrieve all the POIs found inside the buffer area by their name (*rdf:label*), its geographical position.
- *Search of POIs on streets.* We use two spatial functions (Intersect and Buffer), the only necessary parameter is a geographic position. This may be the current location of the user, POI position or an arbitrary position. These queries use the street map and POIs RDF representation. The first query asks for the nearest street from the given position, once the street name are retrieved, we proceed to make another query, asking for all segments that comprise the street, which all the segment define a distance ratio for a buffer that will be implemented on each segment for retrieving all nearly POIs.

4 Experimental Results

4.1 Simple Searches

For these operations, the system retrieves all the name of instances (search by name) or category (search by category) or activity (search by category) in the ontology network. For the search by the name, the results fill out an array that is used to perform an autofill control. When user types a name, the system displays a list with all the POIs that were combined according to the typed letters. In our case, we select "Liverpool" from the list; this POI will be displayed in the map as a landmark (see Fig. 4a). This POI has a disclosure button, which displays the details of the POI from our information and retrieved data from repositories. If it exists, then a link between is performed (see Fig. 4b). In the case of categories, they show all the categories (see Fig. 4c) and when they are selected for displaying, the query for all instances that belong to the selected categories are visualized in the map (see Fig. 4d).

Fig. 4 Simple search cases **a** selected POI displayed on the map, **b** POI description, **c** existing categories of POIs, **d** POIs are displayed by their Category

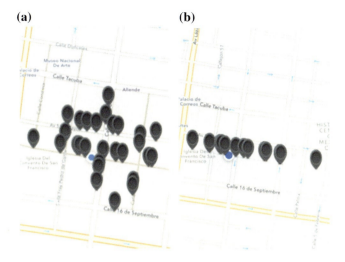

Fig. 5 Spatial operations cases **a** set of POIs into a search buffer, **b** set of POIs on a selected street

4.2 Spatial Operations

For the spatial operations, we use the user current location for performing them. First for searching by distance ratio, the system displays a bar for the user, in which the value is set to meters. This query returns all POIs inside the search ratio (see Fig. 5a). For searching by street, it is necessary to choose the intersect button to perform this operation. In this case, we ask for all the POIs on the "Francisco I. Madero" street (see Fig. 5b).

5 Conclusions and Further Work

In this work, we propose an approach that is capable to generate personalized touristic itineraries using a user profile and weighted features for performing recommendations about POIs. The approach is also capable to retrieve information from web, using SPARQL queries to endpoints from external data sources by implementing the Linked Data principles.

On the other hand, according to the POI information, we transform the original data (shapefiles) into RDF files in order to share and integrate information. So, it can populate our ontology network. This allows us to store the ontology network into a triple store, creating a SPARQL endpoint for querying and performing spatial operations. The process removes the use of traditional spatial databases. When the Linked Data principles are implemented, all POIs could be enriched by information from repositories where the POI exists with different information, which is complement of the POI detail.

Future work we will implement the OGC standard GeoSPARQL, this for give to system the capability of perform spatial operations on a SPARQL Endpoint instead use a spatial extension of a DataBase Management System (DBMS), also we will develop a mobile application, in this case for Apple devices, with the purpose to proof these methods in a real environment with real users, besides we will generate a personalized itineraries for each user using all the described methods in this paper.

Acknowledgments This work was partially sponsored by the IPN, CONACYT and SIP, under grant 20140545. Additionally, we are thankful to the reviewers for their invaluable and constructive feedback that helped improve the quality of the paper.

References

1. Schiaffino S, Amandi A (2009) Intelligent user profiling. In: Artificial intelligence an international perspective, Springer, Berlin, pp 193–216
2. Bobadilla J, Ortega F, Hernando A, Gutiérrez A (2013) Recommender systems survey. Knowl Based Syst 46:109–132
3. Sarwar B, Karypis G, Konstan J, Riedl J (2001) Item-based collaborative filtering recommendation algorithms. In: Proceedings of the 10th international conference on World Wide Web. ACM, pp 285–295
4. Zibuschka J, Rannenberg K, Kölsch T (2011) Location-based services. In: Digital privacy, Springer, Berlin, pp 679–695
5. Ruotsalo T, Haav K, Stoyanov A, Roche S, Fani E, Deliai R, Mäkelä E, Kauppinen T, Hyvönen E (2013) SMARTMUSEUM: a mobile recommender system for the web of data. Web Semant Sci Serv Agents World Wide Web 20:50–67
6. García A, Chamizo J, Rivera I, Mencke M, Colomo R, Gómez JM (2009) SPETA: social pervasive e-tourism advisor. Telematics Inform 26(3):306–315
7. Moreno A, Valls A, Isern D, Marin L, Borràs J (2013) SigTur/E-Destination: ontology-based personalized recommendation of tourism and leisure activities. Eng Appl Artif Intell 26 (1):633–651
8. Becker C, Bizer C (2008) DBpedia mobile: a location-enabled linked data browser, vol 369. LDOW, Beijing
9. Mao Y, Peifeng Y, Wang-Chien L (2010) Location recommendation for location-based social networks. In: Proceedings of the 18th SIGSPATIAL international conference on advances in geographic information systems (GIS 2010), New York, NY, pp 458–461
10. Middleton SE, Roure DD, Shadbolt NR (2009) Ontology-based recommender systems. In: Staab S, Studer R (eds) Handbook on ontologies, international handbooks information system. Springer, Berlin, pp 779–796
11. Gruber T (1995) Towards principles for the design of ontologies used for knowledge sharing. Int J Hum-Comput Stud 43(5/6):907–928
12. Golemati M, Katifori A, Vassilakis C, Lepouras G, Halatsis C (2007) Creating an ontology for the user profile: method and applications, In: First IEEE international conference on research challenges in information science (RCIS), Morocco
13. Luna V et al (2014) An ontology-based approach for representing the interaction process between user profile and its context for collaborative learning environments. Comput Hum Behav 21:623–643
14. Buriano L, Marchetti M, Carmagnola F, Cena F, Gena C, Torre I (2006) The role of ontologies in context-aware recommender systems. In: 7th international conference on mobile data management, MDM 2006, pp 80, 10–12 May 2006

15. Rich E (1983) Users are individuals: individualizing user models. Int J Man-Mach Stud 18(3):199–214
16. Sarwar BM, Karypis G, Konstan J, Riedl J (2002) Recommender systems for large-scale e-commerce: scalable neighborhood formation using clustering. In Proceedings of the fifth international conference on computer and information technology, vol 1, pp 5–32
17. Linden G, Smith B, York J, (2003) Amazon.com recommendations: item-to-item collaborative filtering, IEEE Internet Comput 7(1):76–80
18. Adomavicius G, Tuzhilin A (2005) Toward the next generation of recommender systems: a survey of the state-of-the-art and possible extensions. IEEE Trans Knowl Data Eng 17(6):734–749
19. Candillier L, Meyer F, Boullé M (2007) Comparing state-of-the-art collaborative filtering systems. Lect Notes Comput Sci 4571:548–562
20. Schafer JB, Frankowski D, Herlocker J, Sen S (2007) Collaborative filltering recommender systems. Adapt Web 9:291–324
21. Vilches-Blázquez LM (2011) Metodología para la integración basada en ontologías de información de bases de datos heterogéneas en el dominio hidrográfico (Ph.D. thesis Universidad Politécnica de Madrid)
22. Fonseca F, Câmara G, Monteiro AM (2006) A framework for measuring the interoperability of geo-ontologies. Spat Cogn Comput 6(4):307–329
23. Ressler J, Dean M (2007) Geospatial ontology trade study. In: Ontology for the Intelligence Community (OIC-2007), November 28–29, Columbia, Maryland
24. Höepken W, Clissmann H (2006) Harmo-TEN tourism harmonisation trans-European network, vol 3. Retrieved from www.harmo-ten.info/harmoten_docs/D2_2_Ontology_User_Manual_
25. Huang Y, Bian L (2009) A Bayesian network and analytic hierarchy process based personalized recommendations for tourist attractions over the internet. Expert Syst Appl 36(1):933–943
26. Allocca C, D'Aquin M, Motta E (2009) DOOR—towards a formalization of ontology relations. In Dietz JLG (ed) KEOD, pp 13–20
27. Suárez MC, Gómez A (2012) The NeOn methodology for ontology engineering. In: Ontology engineering in a networked world, Springer, Berlin, pp 9–34

Part III
Geospatial Crowdsourcing, Modeling and Computing

Detecting Clustering Scales with the Incremental K-Function: Comparison Tests on Actual and Simulated Geospatial Datasets

Ran Tao, Jean-Claude Thill and Ikuho Yamada

Abstract The detection of so-called hot-spots in point datasets is important to generalize the spatial structures and properties in geospatial datasets. This is all the more important when spatial big data analytics is concerned. The K-function is regarded as one of the most effective methods to detect departures from randomness, high concentrations of point events and to examine the scale properties of a spatial point pattern. However, when applied to a pattern exhibiting local clusters, it can hardly determine the true scales of an observed pattern. We use a variant of the K-function that examines the number of events within a particular distance increment rather than the total number of events within a distance range. We compare the Incremental K-function to the standard K-function in terms of its fundamental properties and demonstrate the differences using several simulated point processes, which allow us to explore the range of conditions under which differences are obtained, as well as on a real-world geospatial dataset.

Keywords K-function · Point data · Hot spots · Spatial clustering · Scale · Spatial big data

---

R. Tao · J.-C. Thill (✉)
Department of Geography and Earth Sciences and Project Mosaic,
University of North Carolina at Charlotte, Charlotte, NC, USA
e-mail: Jean-Claude.Thill@uncc.edu

R. Tao
e-mail: rtao2@uncc.edu

I. Yamada
Department of Integrated Science and Engineering for Sustainable Society,
Chuo University, Tokyo, Japan
e-mail: iyamada.87@g.chuo-u.ac.jp

© Springer International Publishing Switzerland 2015
V. Popovich et al. (eds.), *Information Fusion and Geographic Information Systems (IF&GIS' 2015)*, Lecture Notes in Geoinformation and Cartography,
DOI 10.1007/978-3-319-16667-4_6

1 Introduction

The analysis of spatial point patterns has a long tradition in various scientific disciplines including geography, economics, demography, ecology, forestry, criminology, epidemiology, planning, and business [1, 2]. By detecting and analyzing the spatial point patterns, something interesting and informative about the underlying process that would have generated the events can be unveiled. For example, studies of animal behavior suggest that agglomerative or clustering point patterns are helpful to verify theories of territoriality and social organization; also, a diffusion point process in both spatial and temporal dimensions can provide evidence for various theories about information transmission or disease spreading [3]. A spatial concentration of cell phone users with certain attributes within a narrow time window may present a business opportunity or a risk to public safety.

Spatial point pattern analysis has recently regained popularity in spatial sciences and affiliated disciplines as an increasingly large volume of thematically diverse geospatial data is available as point data, with their own x-y coordinates, often with a time stamp. Developments in spatial technologies such as location-aware and remote sensing, advancements in information and communication technologies, along with data sharing inclination by public organizations and individuals in the form of social networking services and other public participation initiatives have created a data rich environment for social sciences that has no precedent in human history. Volunteered Geographic Information (VGI) has become a legitimate complement of existing data sets, which are still often published only at a spatially aggregated level for confidentiality reasons. The importance of this so-called "big data" defined with V criteria [4] has been well recognized by scientists across disciplines. In this "data avalanche" revolution [5], spatial sciences have a unique and vital role to play as most of the big data are georeferenced.

This paper is a contribution to the body of literature aimed at determining the existence of patterns in geospatial point datasets, particularly clumping or clustering structures. We propose a variant on the statistics based on the well-known K-function that would avoid overstating the spatial scope of clusters that may be detected in the empirical datasets. The so-called Incremental K-function is presented and evidence of its efficacy on real data and multiple simulated data sets are presented.

The rest of this paper is organized as follows. In the second part, we summarize the literature on spatial point pattern analysis. Next, we present the Incremental K-function method and its important properties. Then we conduct the comparison experiments with the conventional K-function on both real-world datasets and simulated ones and analyze the results. In the final part, we conclude on the characteristics and practical usefulness of the Incremental K-function.

2 Spatial Point Pattern Analysis

As one of the most common spatial patterns due to the general tendency of spatial phenomena (i.e. events) to co-occur spatially as encapsulated by Tobler's First Law of Geography [6], spatial clustering represents a general tendency of events occurring closer to each other than one might expect and it always draws great attentions in academics [7]. Clusters or clumps of events in the geographic space are commonly called hot spots. The detection of hot spots in one-dimensional spatial big data sets is of significance because it serves to generalize data and their spatial properties, which is critical for inferential purposes. A significant body of research has contributed to developing methodologies and tool sets for detecting spatial clustering. In the context of point pattern analysis, this family of methods is named second-order analysis of point processes [8, 9], or hot spots detection.

Early studies were primarily concerned with the overall pattern embedded in the spatial events. Therefore, a number of single-index spatial statistics, sometimes labeled as "global" statistics, were designed to depict the nature of events within the entire study area. Well-known examples include Moran's I, Geary's C, Quadrat Analysis, Nearest Neighbor Index, G statistic, and Ripley's K-function. Later on, scholars found that one of the fundamental assumptions of the global statistics, namely the spatial stationarity, is difficult to be held in many real situations. In addition, with only a single statistic to describe the entire study area, it is inadequate to further investigate more detailed aspects such as how the distribution of one variable would affect another in a localized fashion, or where departures from a random spatial distribution can be found [10]. To address these issues, along with the fast development of GIS in the 1990s, the study trend shifted to developing local statistics of detecting spatial clusters, i.e. 'hot spots'. Noticeable approaches include the geographical analysis machine (GAM) [11] and its derivative methods [1, 12], the local version of Ripley's K-function [9], local indicators of spatial association (LISA) especially the local Moran's I statistic, local Geary's C [13], and local G statistic [14, 15]. In contrast with their global counterparts, the local techniques are aimed at finding anomalies and interesting collections of spatial events within the study area that appear to be inconsistent with the background conceptual model of how events arise or at pinpointing the specific locations that serve as foci for clustering that repeats itself over the study area [7, 12]. Sometimes local methods with predetermined locations are given a special name "focused tests" to differentiate them with the ones based on randomly-chosen event locations [12]. More recently, related studies are aimed at handling emerging large point datasets with accurate locational information, sometimes with a time stamp. Techniques of Exploratory Spatial Data Analysis (ESDA), GeoComputation, and GeoVisualization are frequently incorporated for this purpose [16–18].

Among various methods of point pattern analysis, Ripley's K-function is regarded as one of the most effective methods at detecting whether a spatial process significantly departs from randomness, or whether it is more dispersed or more concentrated than random. The K-function is routinely used as a technique for hot spots detection, that is for the discovery of high concentrations of point events. It has been enhanced through the decades since it was originally proposed by Ripley in 1976 [8, 19]. The fundamental idea of the K-function is to count the number of events within a certain distance to randomly selected event locations. The number is then used to calculate K-function value by dividing the density of events. To obtain statistical conclusions, the K-function value is compared with the expected value according to the null hypothesis, for example Complete Spatial Randomness (CSR). If the observed value is higher than expected, the study events have the tendency of clustering; or dispersed, if it is lower than expected. Monte Carlo simulation is frequently applied to assess the statistical significance of the results [11]. Several meaningful extensions and applications of the K-function have been conducted through the years. Of note is work by Getis and Franklin [9], where spatially local clusters are detected within the range of K-function, which is subsequently known as local K-function analysis. Boots and Okabe [20] discussed applying the Cross K-function as a focused test to identify clusters of events around specific locations for example crime cases surrounding rail stations. Yamada and Thill [2] adjusted both the global and local K-function to network-constrained space to study transportation-related cases. Tang et al. [18] incorporated Graphics Processing Units (GPU) technique to accelerate the computing process of Ripley's K Function for massive spatial point datasets.

Compared with other hot spot detection techniques, the K-function holds a unique advantage that spatial dependence is examined over a range of distances and spatial scales. For instance to analyze the spatial pattern of crime events within the city, with the K-function we can set the detecting radius from 0.1 to 10 miles so that clusters of the size within this range can all be discovered. This advantage enables the K-function to compare point patterns across scales and easily pick the most interesting ones without the painful process of choosing appropriate spatial weight matrix for LISA or deciding proper size and shape for quadrats. This is the reason the K-function is also called 'Multi-Distance Spatial Cluster Analysis' in the literature.

However, determining the exact scale of an observed pattern from the multiple scales examined by K-function remains problematic. In fact, most of the existing application studies simply choose an "optimal" scale to show the results without further explanation. Moreover, when applied to a pattern exhibiting local clusters, the observed K-function tends to exceed the upper significance envelope even at scales beyond the process's true scale [2]. As a result, the point patterns detected by K-function would be untrue at certain scales and further inferential conclusions built on this would be misleading as well. In order to remedy this issue, we use a variant of the K-function namely the Incremental K-function in this paper. Instead of counting the total number of events within a distance range, the Incremental K-function examines the number of events within a particular increment of distance. We will further discuss the fundamental properties of the Incremental K-function in the next section, compare them to those of the standard K-function and demonstrate the differences on several real-world geospatial datasets as well as simulated point processes. The purpose is to explore the capability of the Incremental K-function that complements the K-function methodology family for detecting point patterns at the spatial process's true scale.

3 Incremental K-Function

The Incremental K-function is a variant of the standard K-function. To differentiate with their local versions, the standard K-function and the Incremental K-function is also referred to as the global K-function and the global Incremental K-function in this paper. Let us consider a point process including n events $P = \{p_1, p_2, \ldots, p_n\}$ in a study region, the global K-function is defined as

$$K(r) = \frac{1}{\rho} E(\text{number of event in P within distance } r \text{ of an arbitrary event of P}) \quad (1)$$

here ρ is the density of events within the study region. It can be further written as:

$$K(r) = \frac{A}{n(n-1)} \sum_{i=1}^{n} \sum_{j=1, j \neq i}^{n} I_{i,j} \quad (2)$$

where r is the detection window radius or scale; A represents the total area of the study region; n is the total number of events in the study region; $I_{i,j}$ equals to one if the distance between event i and event j is less than r, and zero otherwise.

By decomposing the global K-function down to an individual event location i we can obtain the local K-function [9] as:

$$LocK_i(r) = E(\text{number of events in P within distance } r \text{ of event } i) \quad (3)$$

Or it can be formulized as:

$$LocK_i(r) = \sum_{i=1}^{n} I_{i,j} \quad (4)$$

In contrast, the Incremental K-function counts the number of events within a particular increment of distance, i.e. the "donut" area from the smaller scale r_{t-1} to the current scale r_t. The only exception is when r_t is the smallest scale r_1 the Incremental K-function is then the same as the K-function. The formula of the global Incremental K-function is defined as:

$$IncK(r_t) = \begin{cases} K(r_t) - K(r_{t-1}), & t = 2, 3, \ldots, \\ K(r_t), & t = 1 \end{cases} \quad (5)$$

And the local version of the Incremental K-function is defined by the same logic:

$$LocIncK_i(r_t) = \begin{cases} LocK_i(r_t) - LocK_i(r_{t-1}), & t = 2, 3, \ldots, \\ LocK_i(r_t), & t = 1 \end{cases} \quad (6)$$

Fig. 1 Example of (incremental) K-function

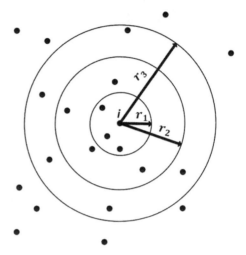

Taking the example shown in Fig. 1, and assuming a unit density of events, the local K-function value for event location i equals 3 at the scale of r_1; 7 at the scale of r_2; and 12 at the scale of r_3, while the local Incremental K-function value equals to 3, 4, and 5 at scale of r_1, r_2, and r_3 respectively. The global K-function and the global Incremental K-function will also result differently as they are comprised of their local counterparts. As illustrated by Yamada and Thill [2], a false positive error may be associated with the observed K-function of a pattern exhibiting a local clustering tendency at certain scales where the Incremental K-function will truthfully fail to detect such tendency. Because the K-function averages densities over the entire distance range, it may overshoot the true cluster size.

4 Comparison Experiments

In this section we implement a series of comparison experiments with both data simulated to represent known spatial processes and real-world geospatial data. The global and local K function as well as the global and local Incremental K-function values are calculated according to the equations given in the previous section. Statistical inference is determined by 1,000-time Monte-Carlo simulation based on the assumption of Complete Spatial Randomness (CSR) of the point process. We adopt the significance level of 5 % for global results and 0.1 % for local results to account for test simultaneity. Edge effects are corrected through shrinking the size of the analysis area by a distance equal to the largest distance band to be used in the analysis. Only the points within the shrunk area are used to calculate K-function and Incremental K-function values, while the background point process and the simulated points remain within the original area. The implementation program is coded in C/C++ and computed with a 64 bit PC with 4 CPUs and 16 GB RAM. The parallel computing technique OpenMP is applied to accelerate the computation process, which is particularly beneficial for the Monte Carlo simulation task.

We first conduct tests with simulated datasets. In a square region of 10,000 by 10,000 units, we simulate a series of point patterns generated according to different known processes. The total number of points is 2,500 for every point pattern. We set our detection scale from 100 units to as many as 2,500 units in order to scan the full extent of point patterns. In order to verify our approach, we start by simulating a random point pattern based on the null hypothesis namely Complete Spatial Randomness, which is used as a benchmark (as shown in Fig. 2). On the right-hand side of Fig. 2, we show the global K-function results on top and the global Incremental K-function results at the bottom. The blue line represents the value of either K-function or Incremental K-function, while the red and green lines stand for statistical significance envelopes numerically simulated through Monte Carlo process. Not surprisingly, in both charts of Fig. 2 the three lines closely overlap with each other. It indicates that neither the K-function nor the Incremental K-function of this random point pattern has escaped the expected range generated by CSR. In other words the two functions both have successfully verified the random point pattern.

We are more interested in the capabilities of these two functions to deal with clustered point patterns. Figure 3 shows a simulated clustered point patterns in which 25 independent point clusters are generated. Each of the point clusters consists of 100 points distributed according to a Bivariate Normal Distribution with standard deviation of 100, while the centers of these clusters are located on a square

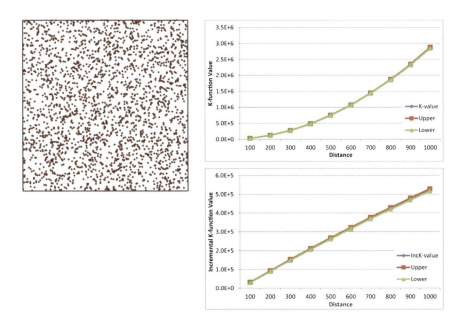

Fig. 2 Random point pattern results

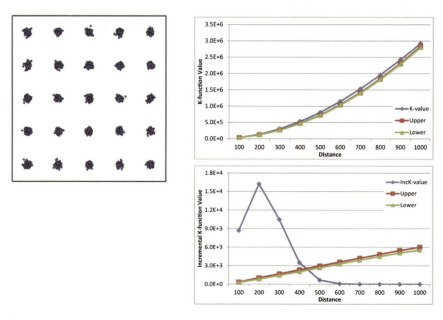

Fig. 3 Clustered point pattern (standard deviation = 100)

grid within the square study area with a distance of 2,000 units. This means that there is a 68 % probability that a point is within 100 units of its local cluster center; this probability becomes 95, 99, and 100 % for distances of 200, 300, and 400 units respectively. The results of the K-function and the Incremental K-function are quite different this time. As we can see from the two charts in Fig. 3, the observed global K-function is above the upper envelope, and thus detects a significant clustering pattern, at most scales. On the contrary, the global Incremental K-function detects a significant clustering pattern only at the first four scales and a significantly regular (or dispersed) pattern at all larger scales. The global incremental K-function has captured the properties of the point process accurately. In the bottom chart of Fig. 3, the two largest increments of 'IncK-value' are at the zero to 100 scale and the 100–200 scale, which correspond to the probability increasing from 0 to 68 % and then to 95 %. Beyond the 200-unit scale, the increment of the function value starts dropping as there only remains less than 5 % probability to include more points into a local cluster. Beyond the 400-unit scale, the probability drops down to zero. Correspondingly, we observe the 'IncK-value' in the figure drops below the lower significance envelope. In short, the incremental K-function varies consistently with the underlying point process. At the small scales the 'IncK-value' stays beyond the upper significance envelope indicating the local clusters dominate the global point pattern of the whole study area. With the scale increasing beyond a threshold between 400 and 500 units, the local clusters become more like single points and their grid-like centers have swayed the global point pattern to a regular one. Evidence can also be found in the results of the Incremental K-function at larger scales. Overall, the K-function has failed to detect this changing point process. The clustered point pattern at almost all the scales indicated by the K-function is definitely inconsistent with our simulated point process. We also conduct a similar test in which only the size of local clusters has changed (to a standard deviation of 300). Figure 4 illustrates this point pattern and its two comparison results. Again the Incremental K-function has captured the nature of the point process across scales. It indicates a clustered pattern at small scales and a regular one at larger scales. The only difference with the results in Fig. 3 is the peak is shifted toward larger scales, which corresponds well to the enlarged local clusters. In contrast, the K-function result shows a clustered pattern across all scales, which would clearly be misleading at large scales.

To fully unveil the capability and characteristics of the Incremental K-function, we also experiment on more complex datasets. Figure 5 shows a group test on a two-stage clustered point dataset. The basic features of this data remain the same as the one in Fig. 4 in terms of the number of clusters as well as the size of each cluster (standard deviation of bi-normal distribution is still 300). However the center of several clusters is relocated so as to have four larger clusters in the four corners, each one being former of smaller clusters (there is still a total of 25 small clusters). Comparing the results of the two global functions, we find extreme differences again. The K-function once again results in clustered patterns regardless of the scale that is varied from 100 to 2,500 units, from which we can hardly extract any more useful information. By contrast, the results of Incremental K-function coincide with the underlying two-stage clustered point process. It indicates two and only two

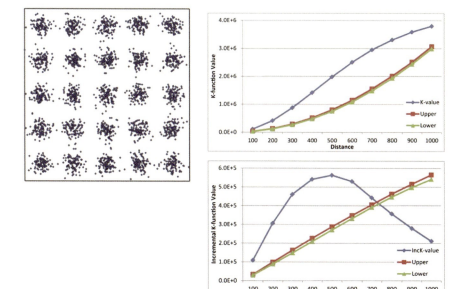

Fig. 4 Clustered point pattern (standard deviation = 300)

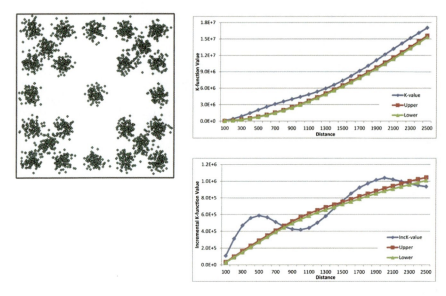

Fig. 5 Two-stage clustered point pattern (standard deviation = 300)

separate peaks of significant clustering. The first peak shows up at the same scale (500 units) as Fig. 4 because the size of local clusters remains the same. The second peak manifests itself at the scale of 2000 and demonstrates that the Incremental K-function is capable to detect those larger clusters formed of local clusters. Therefore it does neither overestimate nor underestimate the point patterns. The conclusion from this group of experiments is that the Incremental K-function can not only detect out at which scale it shows clustering, but is also able to capture the variation of cluster size across scales.

Furthermore, we carry out experiments on even more complex datasets with various sizes of clusters at random locations. The geospatial dataset depicted in Fig. 6 includes 25 local clusters, of which 10, 8, and 7 clusters are generated based on a bi-normal distribution with standard deviation of 100, 300, and 500, respectively. The results are consistent with the previous experiment. On the one hand, the K-function detects significant clustering at all scales ranging from 100 to 2000, thus committing patent false positive error. On the other hand, three separate peaks of clustering pattern corresponding to the three scales used to generate the point distributions are picked out by the Incremental K-function in spite of the random location of cluster centers. It provides further evidence of the Incremental K-function's sensitivity and accuracy with respect to point patterns across scales.

As a matter of fact, real-world situations are usually more complex than simulated ones. Therefore we have also implemented comparison studies with real-world geospatial data to backup and supplement the conclusions obtained from simulated data. The data we use in this study are the records of vehicle theft and recovery locations in Charlotte, North Carolina. Given the extremely heterogeneous

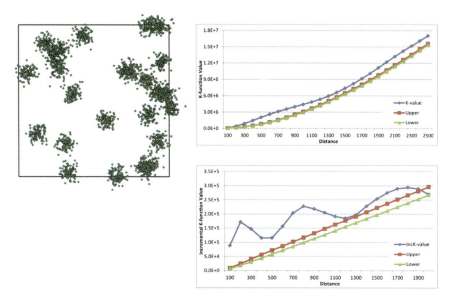

Fig. 6 Clustered point pattern with random seeds and various cluster scales

distribution of these records, the study area is narrowed down to an eight-mile by eight-mile square region surrounding the city center. According to the crime report database maintained by the Charlotte-Mecklenburg Police Department (CMPD), there were 1,832 vehicles stolen within this region from January 1st, 2012 to December 31st, 2013, of which 995 have been recovered somewhere else in the city. Recovery locations show a more varied point pattern across scales than the theft locations. Therefore we use these 995 vehicle recovery locations to illustrate the properties of the Incremental K-function and explore its practical usefulness as well.

Figure 7 shows the study region and the auto recovery locations for the stated period of analysis. The right-hand side panels present the global K-function and the global Incremental K-function charts. The smallest detection scale is set at 0.05 mile, i.e. 80 m, and 20 times that much as the largest possible scale. The K-function shows significant clustering over the entire range of scales as the 'K-value' stays well above the upper significance envelope and smoothly departs from it to indicate an even more significantly clustered pattern for larger scales. On the other hand, the 'IncK-value' in the bottom chart is quite uneven across scales, although it stays above the upper envelope at all scales. Its several up-and-down periods are believed to reflect the heterogeneous nature of this real-world dataset. A first peak shows up at the smallest possible scale and also exhibits the greatest departure from the significance envelopes. From this, we can conclude that there exist a number of auto recovery location clusters across the study region and that these clusters have very compact sizes as their radii are 80 m or less. Through further investigation we found these small clusters match specific geographical features such as car dismantlers or deserted parking lots where cars would be abandoned. Besides the global maximum

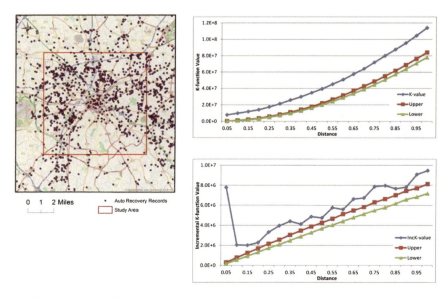

Fig. 7 Global results of Charlotte auto-recovery data

at the 0.05 scale, other local maxima in the Incremental K-function are found at 0.35 and 0.8 mile, as well as several less prominent ones. These could be explained as thieves' favorite car dumping areas, for example unsupervised neighborhoods or the vicinity of the airport.

Beside the analysis of the global spatial pattern in the car recovery locations, we also apply the local version of two K-functions to investigate clusters locally. We implement the local functions following the same process as the global ones, but change the significance level to 0.1 %. While the local analysis can be conducted at any and all discrete location in the study region, for the sake of the illustration we pick out one specific record located to the Northeast of the city center. Figure 8 includes a map of this localized area and the plots of two local K-functions. This time the local Incremental K-function returns two significant peaks with one at the 0.05 mile scale and the other at 0.75 mile. On the map at left-hand side, the scales of 0.05, 0.1, 0.7, and 0.75 mile are highlighted using red circles. The smallest circle encompasses 17 events, while the second one includes no additional one. Comparing the results of two local K-functions, the 'Local IncK-value' has captured this situation rather crisply as it indicates clustering only at the first scale. Conversely, the 'Local K-value' shows clustering at both of the first two scales and even beyond (although at a decreasingly level of significance), which is a clear case of false positive error and gives a misleading message that the local cluster keeps enlarging. The ring-like area between 0.7 and 0.75 mile scale includes 13 neighboring events and it results in another significant peak in the 'local IncK-value'. Examination of the map of events around the selected focal point suggests that no clustering tendency is detectable between these two scales; this is corroborated by

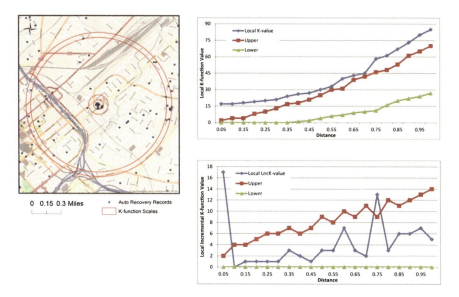

Fig. 8 Local results of Charlotte auto-recovery data

the local Incremental K-function, while the local K-function overdetects a clustering tendency. By looking at the global and local results side by side, it is clear that local point patterns have contribution to the global pattern. Both the global and local Incremental K-functions are capable to reflect the heterogeneous point process and their results can certainly give more meaningful guidance than the original K-functions.

5 Conclusions

The K-function is regarded as one of the most effective methods to detect nonrandom tendencies in a geospatial distribution of points and particularly high concentrations of point events. Although it is designed as a 'Multi-Distance Spatial Cluster Analysis', we argue that it cannot reflect the cross-scale changes of point pattern very well. Instead, we bring up again one of its variant, namely the Incremental K-function, as a solution to this problem. In this paper, we presented the results of a series of comparison studies on both simulated datasets and a realworld dataset. The results from simulated datasets indicate that the Incremental K-function can pick out the scales at which it shows most significantly clustering patterns without the false positive errors that can be so pervasive with the K-function. These peak scales coincide with the true scales designed in our data simulation processes. In addition, the Incremental K-function is also capable to deal with more complex situations such as two-stage clusters and random located clusters of various scales. Meaningful information about how the point pattern varying across scales can be extracted. In contrast, the standard K-function can only offer us a coarse clustering pattern from scale to scale. Given the controlled conditions of the experiments done on simulated point patterns with known properties, it has been demonstrated that the K-function is afflicted by false positive error flaws and incapable of capturing the true scale of point processes.

Moreover, the results from the Charlotte vehicle recovery dataset provide real-world evidence that the Incremental K-function can accurately reflect the underlying heterogeneous point process, and that it does so more reliably that the K-function. Practical usefulness can also be obtained. For instance, the very compact cluster size (80 m) directs that there exist some individual facilities or locations hat concentrate a significant number of stolen vehicles. The local Incremental K-function can serve to pinpoint these locations on the city map for further investigation. To conclude, we hope this work can bring the Incremental K-function to scholars' attention as an effective hot-spot detection method especially dealing with complex spatial point processes.

References

1. Fotheringham AS, Zhan FB (1996) A comparison of three exploratory methods for cluster detection in spatial point patterns. Geogr Anal 28(3):200–218
2. Yamada I, Thill J-C (2007) Local indicators of network-constrained clusters in spatial point patterns. Geogr Anal 39(3):268–292
3. Getis A, Boots BN (1978) Models of spatial processes: an approach to the study of point, line and area patterns, vol 198. Cambridge University Press, Cambridge
4. Laney D (2001) 3D data management: controlling data volume, velocity and variety. META Group Research Note, 6
5. Miller HJ (2010) The data avalanche is here. Shouldn't we be digging? J Reg Sci 50 (1):181–201
6. Tobler WR (1970) A computer movie simulating urban growth in the Detroit region. Econ Geogr 46(2):234–240
7. Waller LA (2009) Detection of clustering in spatial data. The Sage Handbook of Spatial Analysis. London, pp 299–320
8. Ripley BD (1976) The second-order analysis of stationary point processes. J Appl Probab 13:255–266
9. Getis A, Franklin J (1987) Second-order neighborhood analysis of mapped point patterns. Ecology 68:473–477
10. Fotheringham AS (1997) Trends in quantitative methods I: stressing the local. Prog Hum Geogr 21(1):88–96
11. Openshaw S, Charlton M, Wymer C, Craft A (1987) A mark 1 geographical analysis machine for the automated analysis of point data sets. Int J Geogr Inf Syst 1(4):335–358
12. Besag J, Newell J (1991) The detection of clusters in rare diseases. J Roy Stat Soc A (Stat Soc) 154:143–155
13. Anselin L (1995) Local indicators of spatial association—LISA. Geogr Anal 27(2):93–115
14. Getis A, Ord JK (1992) The analysis of spatial association by use of distance statistics. Geogr Anal 24(3):189–206

15. Ord JK, Getis A (1995) Local spatial autocorrelation statistics: distributional issues and an application. Geogr Anal 27(4):286–306
16. Aldstadt J, Getis A (2006) Using AMOEBA to create a spatial weights matrix and identify spatial clusters. Geogr Anal 38(4):327–343
17. Widener MJ, Crago NC, Aldstadt J (2012) Developing a parallel computational implementation of AMOEBA. Int J Geogr Inf Sci 26(9):1707–1723
18. Tang W, Feng W, Jia M (2014) Massively parallel spatial point pattern analysis: Ripley's K function accelerated using graphics processing units. Int J Geogr Inf Sci 28(5):1107–1127 (Accepted)
19. Okabe A, Boots B, Satoh T (2010) A class of local and global k functions and their exact statistical methods. In: Perspectives on spatial data analysis. Springer, Berlin, pp 101–112
20. Boots B, Okabe A (2007) Local statistical spatial analysis: inventory and prospect. Int J Geogr Inf Sci 21(4):355–375

Crowd Computing Framework for Geoinformation Tasks

Alexander Smirnov and Andrew Ponomarev

Abstract In the paper a general purpose crowd computing framework architecture is discussed. The proposed framework can be used to compose crowd computing workflows of different complexity. Its prominent features include ontological description of crowd members' competencies profiles; automatic assignment of tasks to crowd members; the support of both human and non-human computing units (hybrid crowd); and spatial features of crowd members which make way for employing the proposed framework for a variety of crowdsourced geoinformation tasks.

Keywords Crowd computing · Crowdsourcing · Ontology · Profiling

1 Introduction

Geoinformation technologies include a wide spectrum of tasks, related to collection, processing and presentation of geospatial information. Most of these tasks permit automation and, therefore, are solved mostly in automated way in modern geoinformation systems. However, some problems are hard to deal with solely by automated computer environments. The reasons for that may be of different nature. Probably, the most frequently discussed one is connected with the difficulty that typical information processing techniques have dealing with problems involving heavy usage of common sense knowledge and incomplete definitions. Another

---

A. Smirnov (✉) · A. Ponomarev
St.Petersburg Institute for Informatics and Automation of the Russian Academy of Sciences, 39, 14th Liniya, St.Petersburg 199178, Russia
e-mail: smir@iias.spb.su

A. Ponomarev
e-mail: ponomarev@iias.spb.su

A. Smirnov · A. Ponomarev
ITMO University, 49, Kronverksky Pr., St.Petersburg 197101, Russia

© Springer International Publishing Switzerland 2015
V. Popovich et al. (eds.), *Information Fusion and Geographic Information Systems (IF&GIS' 2015)*, Lecture Notes in Geoinformation and Cartography,
DOI 10.1007/978-3-319-16667-4_7

reason, less important but still valid, is physical absence of autonomous sensing/computing devices in places where they are needed to be.

This results in a new kind geoinformation technologies, where classical GIS approaches are amalgamated with social computing practices. Examples of these new kind of technologies are community sense and response systems (e.g., [5]) and crowdsourcing of GIS data. Crowdsourcing becomes more and more widely used source of geospatial information. Examples range from general collaborative mapping (e.g., OpenStreetMap, Google Map Maker, WikiMapia) to thematic projects dedicated to crisis mapping (e.g., [18, 26]), natural resources management [28], e-government [9] and other application areas. Active involvement of non-expert humans into geoinformation processing tasks resulted into emergence of several specific terms, such as *crowd sensing* and *neogeography*.

All these developments are closely related to even bigger research direction aimed on the creation of hybrid human-computer systems, where human does not always consumes the results that are provided by computer devices, but can also be a provider of some information or service that is consumed by other humans or computer devices. From the point of this research direction the classical dichotomy represented by a human that provides some aims and inputs and a machine that performs routine computations to achieve these goals is transformed into a more general perspective of a network of interconnected humans and machines, performing different functions that together achieve the predefined goal.

The special kind of services relying heavily on human-specific abilities are currently discussed under a set of names and in several closely related research areas. The most prominent are crowdsourcing, human computations and crowd computing. All these areas have much in common. The interconnection between them is discussed, for example, in [8]. In this paper, crowd computing is understood as a spectrum of methods and technologies to solve problems with the help of undefined and generally large group of people, communicated through the internet (crowd). This practical definition used by the authors matches well with the specific characteristics of crowd computing that were enumerated in [21] after literature analysis.

The contribution of this paper is twofold. First, it presents a general purpose crowd computing framework that can be used for variety of problems. Second, it shows how this framework can be used to employ crowd computing for problems involving the processing of spatially enabled data.

One of the tasks performed by the authors of this work is the decomposition of currently established practice of programming for hybrid computer-human environments to identify the set of primitive operations that can be used to construct any type of the information processing workflow. In some sense, this task is similar to designing a set of machine instructions, a composition of which allows building any computer program (for hardware computer).

This paper is organized as follows. Section 2 presents current developments in the area of crowd computing framework development. Section 3 discusses the specific features of crowd computing and describes the idea of crowd computing patterns, aimed to form tried-and-tested solutions for typical problems of crowd computing systems. Based on the literature review and specific features analysis, in

Sect. 4 essential requirements are presented that drive the design of the proposed framework. Section 5 presents the overall design of the framework. Section 6 describes how this framework can be used for problems, involving processing of spatially enabled data. Results achieved are summarized in the conclusion.

2 Related Work

This section contains the review of existing multipurpose crowd programming frameworks. The literature analysis has shown that currently there are three major approaches to programming crowd effort: MapReduce-based, workflow-based, database-based.

MapReduce [4] is a programming model for processing and generating large datasets. The primary intent of this model is to allow for programming data processing algorithms easily parallelizable to multiple machines of a cluster. Since its creation in 2004 this programming model has received a lot of attention and has become very popular in the world of distributed data processing.

There were proposed several crowd computing frameworks [1, 10], somehow based on (or inspired by) MapReduce model.

The Jabberwocky programming environment [1] is a set of technologies, including a human and machine resource management system, a parallel programming framework for human and machine computation and a high-level programming language. The programming framework is an adaptation of MapReduce, in which the execution of Map and Reduce functions may require some human actions and the whole process is suspended until these actions are performed. The problem-setter can also impose constraints on characteristics of people who will be asked to perform a task.

CrowdForge [10] is a general-purpose framework for accomplishing complex and interdependent tasks using microtask markets. As a computation model, authors of Crowdforge also adopted MapReduce model augmented by a Partition stage, which purpose is to refine the problem itself and its decomposition into subtasks during the solution process. On the Map stage the problem is decomposed into subtasks which are executed, and on the Reduce stage all the results are merged in a problem-specific way.

Another approach for crowd computing originated in workflow modeling/management systems. The analogy between workflow modeling/management and crowd computing is also quite noticeable: both kinds of systems deal with some process, which can include humans as executive elements. There are several techniques in "pure" business process workflow modeling where human operations are explicitly modeled (BPEL4People, WS-HumanTask). These techniques were generalized and elaborated to achieve a crowd computing programming model.

An example of workflow-based approach to crowd computing is presented, for example, in [11]. The authors propose a three-layered model to crowd computing process. The topmost process/program layer, showing the process context of particular crowd computing task. Crowdsourcing tactic layer, where crowd computing specific operations (such as task parallelizing, quality control) are described. And, the lower-most crowdsourcing operations layer where API calls to some crowdsourcing marketplace are programmed. As a formal scheme for the top two layers the authors use a Business Process Model and Notation 2.0 (BPMN 2.0), proposing an extension to it indicating that some part of the process must be crowdsourced.

In CrowdLang [19, 20] crowd computations are also described in a graphical workflow-based way. The authors propose a set of typical workflow patterns and a graphical notation to specify the computation process. This approach is verified by creating several crowd computing applications for translation from German to English. Another idea that plays an important role in that paper is the focus on reusable crowd programming patterns (represented in the form of workflows).

In papers [6, 16] a database metaphor is developed in the context of crowd programming.

Markus et al. proposed Quirk [16] system and Quirk UDF language. The data model of Quirk is close to relational model supplemented with user defined functions (UDF) for posting tasks and resolving the situation of multiple answers for the same question that can be produced by different crowd members. A whole "program" for crowd in Quirk is represented by a declarative statement very similar to an SQL statement.

Franklin et al. [6] designed a crowd computing system (CrowdDB) in a way inspired by RDBMS. They proposed CrowdSQL, an SQL extension that supports crowd computing. Crowd computing in CrowdSQL is supported both by data definition language (DDL) and data manipulation language (DML) syntaxes. In CrowdSQL DDL, it is possible to define a column or a whole table as crowdsourced, while in CrowdSQL DML a special value similar to SQL NULL value is proposed, meaning that the value should be crowdsourced when it is first used. Beside CrowdSQL language, the authors proposed a user interface generation facility and special query processing techniques for queries involving crowdsourced values.

There are also several *sui generis* crowd computing approaches, that do not fall into one of the categories above.

One of the first multipurpose frameworks for crowd computing is TurKit [13, 15]. It is based on Javascript language with additional library for posting tasks to Amazon Mechanical Turk platform and receiving answers. TurKit uses a "crash-and-rerun" (the name proposed by its authors) programming and execution model designed to suit to long running processes where local computation is cheap, and remote work is costly, which is the case for crowd computing. TurKit suite also includes a generic GUI for running and managing TurKit scripts.

In Turkomatic [12] one of the design goals is to obviate the need for requesters to plan thoroughly through the task decomposition and workflow design. The idea is to crowdsource this information either, i.e. crowd workers are asked to recursively divide complex tasks into simpler ones until they are appropriately short, then to solve them. The requester can also participate and manage the decomposition process and the authors have shown by a case study that requesters involvement usually results in more robust workflows and sane decompositions. Experiments with this approach applied to unstructured tasks (writing an article, vacations planning, composing of simple Java programs) showed that it is effective with respect to certain kinds of tasks.

AutoMan [2] is a domain-specific language and runtime for crowd computing programming. With the use of AutoMan, all the details of crowdsourcing are hidden from the programmer. The AutoMan runtime manages interfacing with the crowdsourcing platform, schedules and determines budgets (both cost and time), and automatically ensures the desired confidence level of the final result. The distinctive feature of this language and runtime is its approach to quality control, based not on the worker features, but on a statistical processing of the results, received from different workers—the system checks whether the results are consistent with the required confidence level and if not, reissues tasks.

3 Human Factors and Crowd Workflow Patterns

One of the most important issues that makes crowd programming significantly special compared with conventional programming for machines is presence of human in the information processing loop. Differences between human and machine in this regard are being analysed on philosophical and technological layers for many decades, but in the context of crowd computing these differences were summarized by Bernstein et al. [3] to the following list:

- Motivational diversity. People, unlike computational systems, require appropriate incentives.
- Cognitive diversity. Characteristics of computer systems—memory, speed, input/output throughput—vary in rather limited range. People, by contrast, vary across many dimensions this implies that we must match tasks to humans based on some expected human characteristics.
- Error diversity. People, unlike computers, are prone to make errors of different nature.

Each of the listed items represents not a particular problem of crowd computing system development, but rather a fundamental issue that results in a bunch of design obstacles and decisions to overcome those obstacles. For example, [2] identifies following challenges for human-based computation:

- Determination of pay and time for tasks. Employers must decide in advance the time allotted to a task and the payment for successful completion.
- Scheduling complexities. Employers must manage the trade-off between latency (humans are relatively slow) and cost (more workers means more money).
- Low quality responses. Human-based computations always need to be checked: worker skills and accuracy vary widely, and they have a financial incentive to minimize their effort. Manual checking does not scale, and simple majority voting is insufficient since workers might agree by random chance.

It is easy to see, that this challenges (addressed in the AutoMan system [2]) are the results of fundamental differences, namely, motivational and error diversity. By the way, cognitive diversity is not addressed in the design of AutoMan.

The first point, taken by the authors of this paper as the basis of our research in the area of crowd computing, is that every crowd computing framework should account for each of these fundamental differences. Moreover, these fundamental differences together provide a basis to describe crowd computing frameworks pointing out the methods and techniques employed to overcome each of these differences. Later, in the respective section the authors will show how the proposed framework addresses all of these differences.

In the rest of this section, the concept of crowd computing patterns is introduced and justified as one of the important concepts in the domain, aimed on dealing with the differences highlighted above.

Like in other branches of computer science and artificial intelligence [7, 24], crowd computing pattern represents a reusable solution to a commonly occurring problem. In crowd computing context this term was first used (to the best of authors' knowledge) by [27] to name various techniques for dealing with unreliability of human responses. Many of these techniques were analyzed and used before that (e.g., [14]), but in [27] there was an attempt performed to describe them as reusable patterns. Later, the concept of crowd computing patterns were investigated in [20] resulting in their own set of reusable workflow constructs. It was also paid some attention in [11], but under the name of "templates" and, moreover, it was stressed there, that crowd computing platform should support reusable process skeletons. The reasons why the idea of crowd computing patterns seems fruitful is twofold. First, these patterns provide tested and effective solutions to error management in human responses. As it was mentioned, error management in this kind of systems is a complex issue, and it does not have an all-fits-one solution: for different kinds of problems different techniques turn out to be most effective. Patterns in this regard help structuring research space of error management and form a body of knowledge about what pattern is favorable in which situations. With this information at hand, a programmer can pick a readily available pattern that shows good results for the concrete problem, or, there is even an opportunity for crowd programming automation. Second, the analysis of this patterns and their building blocks can help to determine the—following an analogy between hardware and crowd computers—instruction set of a crowd computer. A minimal and sufficient set of operations to form any program for a crowd computer.

In [27] three basic patterns were identified:

Divide-and-Conquer. A complex task can be too large for an individual crowd member and, therefore, should require contribution of multiple crowd members. Divide-and-conquer pattern means decomposition of a complex task to several subtasks assigned to different crowd members and followed by composition of subtask solutions received from crowd members into the solution of the initial task.

Redundancy-based quality control. This pattern directly deals with error management in crowd computing. To alleviate the impact of erroneous answers provided by an individual crowd member, one task is assigned to several members and then some quality control mechanism is applied to the answers received to select the most likely correct one. There are several quality control mechanisms developed: averaging, simple voting, weighted voting to name a few. The pattern itself doesn't answer to the question how much redundancy is enough for particular task or what quality control is the best for a certain application, it just forms a scheme of a typical workflow.

Iterative improvement. In this design pattern one task is also assigned to several crowd members, but they work on this task in a sequential manner, and each member is able to see the solution provided by the previous worker. Iterative improvement pattern was successfully applied, for example, for text transcription (e.g., [14]).

An application (for one problem) may be composed of several patterns, for example, the whole problem can be decomposed into several subtask, and each subtask can be distributed to several crowd members with majority voting as the employed quality control mechanism.

4 Requirements

The literature analysis lets to identify a set of requirements for crowd computing framework. This section discusses most important of these requirements, as it is seen by the authors. Hereafter in this section, crowd computing framework will be referred to as "framework" for simplicity.

Framework should provide support of various incentives. The most widely employed in crowd computing incentive schemas are money and reputation, however there are more rare and exotic ones [22].

Framework should provide explicit treatment of cognitive diversity in and between human actors (crowd members). This requirement is a response to one of the fundamental differences between purely machine programming and programming of human-machine systems. In practice, it means that there should be a possibility to describe a crowd member with a set of characteristics corresponding to performance of this member in some type of tasks. The word "possibility" in clarification above should be stressed, because it is an application-specific requirement, and in some applications cognitive diversity can be neglected, but anyway, a multipurpose framework should provide support of it.

Framework should provide abstractions for complex coordination patterns such as quality control or group decision procedures. The advantages of pattern-oriented crowd programming were already enumerated in the respective section of the paper. It is also worth noting, that the list of abstractions provided by the framework should be extensible; in this regard, this requirement is connected to the requirement of providing an abstraction mechanism.

Framework should provide adaptive workflows. Several researchers (e.g., [2, 10, 27]) argue that entire specification of crowd computing in advance is not always the best option. The work specification itself might be adjusted during the time of execution either by human coordinators or by some algorithm and it sometimes leads to better results. This argument is supported by the evidence, so the proposed framework should also support editable/transformable on the fly workflows. However, workflow adaptation requires not only the possibility to change workflow specification in run-time, but also the presence of some entity in the system that implements this adaptation based on the calculation process [27]. Perceived as a part of crowd computing framework, this entity imparts elements of self-organization to it.

Framework should maintain user identities, rich user profiles, and social relationships. All this user information should allow problem setter to select crowd members that should address a task.

Framework should provide an abstraction mechanism, so that users with different skills and with different programming background could organize crowd computing workflow. By no means this prefers simplistic solutions to more advanced, it only requires that there should be several ways to organize workflow ranging from coarse construction from predefined blocks, to fine programming of particular quality control procedures.

Framework should support flexible scheduling. Scheduling in crowd computing is related to the search of balance between time required to solve a problem, dealing with substantial human latency, and cost of using more actors. As it is pinpointed in [2], scheduling in programming systems involving humans represents a very important and difficult task, so the proposed framework should be flexible enough to adopt different scheduling policies. It would allow to experiment with different scheduling models pertaining other components of the systems and code developed for it.

There should be a possibility to add software services to the crowd, resulting in a hybrid crowd. Hybrid crowd [25] is a relatively new research area, which is also developing under different names (e.g., human-machine cloud [17, 23]). The idea behind hybrid crowd reveals conceptual similarity between cloud services and crowd computing, both of which are based on a kind of resource virtualization and pay-per-use basis. Convergence of this two technologies leads to creation of new generation computing environments constituted from humans and machines.

Framework should provide logging and offline analysis facilities. One of the specific features of crowd computing system employing monetary-based incentivization is that execution of some operations cost money. Therefore, it is rather straightforward to try to save as many intermediate results as possible to avoid rerunning expensive operations. On the other hand, the extensive logging of all the intermediate results can provide a basis for applying alternative methods of analysis and to compare different processing methods of one collected dataset.

Framework should be able to use existing crowdsourcing platforms (e.g., Amazon Mechanical Turk), as there these platforms provide not only low-level tools to post task and receive answers, but—the most valuable—these platforms have huge database of users, i.e. provide the access to crowd itself.

5 Framework Design

To meet the requirements identified during the analysis phase, the authors propose a crowd computing framework. Conceptually, the proposed framework consists of three layers:

- The upper layer is *self-organization layer*, which contains components, methods and algorithms for adjusting workflows based on execution process.
- The middle layer (*workflow layer*) performs coordination of crowd computing process based on various patterns of human information workflow. These patterns were identified as a result of literature review: task splitting, different techniques of matching results provided by crowd members, etc.
- The bottom layer (*crowd layer*) contains primitive operations of crowd computer, such as addressing a particular task to a particular crowd member, posting a task into a common task pool, receiving answer from a crowd member.

The two latter layers can be used to easily build various crowd programming workflows, the self-organization layer is in some sense optional, but facilities of this layer can be used to adjust predefined workflow design of a particular application.

Beside three enumerated layers of programmable process, the proposed framework defines metamodels for crowd member and tasks description. These metamodels have very general nature and are equivalent to OWL ontology metamodel.

Runtime infrastructure, provided by the framework, is shown in Fig. 1. Central component of this infrastructure is the *program interpreter* that takes declarative workflow specification, posts human task requests according to this workflow specification and processes the results.

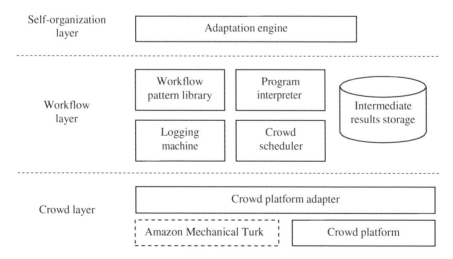

Fig. 1 General view of the framework

Human task requests are then processed by the *crowd scheduler*, which is responsible for assigning tasks to crowd members. Usually, this assignment depends on task attributes, underlying crowd platform features, crowd member declared skills and actual performance. When crowd scheduler finds appropriate members for task requests it posts human tasks to the *crowd platform*.

Crowd platform is the component of the framework responsible to actual communication with crowd members, i.e. sending tasks and receiving answers. One of the goals of this framework design was to adapt to existing crowdsourcing platforms (e.g., Amazon Mechanical Turk), so there is an adapter layer that implements crowd platform interface from one side and uses existing platforms' application programming interfaces from the other side.

Results received from crowd members are documented by the *logging machine*, and then stored in the *intermediate results storage* that holds the global state of all the crowdsourcing process.

The *adaptation engine* analyses the process of computation and transforms running program in order to optimize performance.

The framework is oriented on "top-to-bottom" rewarding calculation for abstract "cost" rewards, excluding other types of incentivization. Initially, a task is assigned some cost limit that the problem-setter can spend on this task. Crowd computing algorithm starts from the whole task and assigns it one unit of resource (corresponding to fund limit). On each task decomposition, resource is divided between subtasks by this algorithm in an application specific way, so that each particular human task is assigned certain cost based on resource value that was obtained after all separations. However, the proposed algorithm fails in situations of dynamic workflows, so there is some room for improvement.

The framework also includes *client module* that is able to communicate with crowd platform. This client module is deployed on crowd member's device and is used to deliver human task to people's devices and collect output. It should be stressed, that client module is able to communicate with original crowd platform only, in case some other crowd platforms are used (e.g. Amazon Mechanical Turk), tasks are assigned by client interfaces of the respective platforms.

Following distinction should be made. The proposed framework presents an infrastructure and generic components for crowd computing applications. An application is created for some particular purpose (collecting the data about traffic jams, prices in supermarkets etc.) with the help of the presented framework's infrastructure. On the other hand, application is to be deployed in a physical machine that also has to have some components (runtime) of the proposed framework, that enable the process of application execution.

An application developed in terms of the proposed framework is defined by the following set of components:

- user model, containing a set of application-specific skills and competencies, represented as an OWL ontology;
- task model, also represented as an OWL ontology;

- declarative code that defines crowd computing algorithm, including user-task matching and reward distribution;
- (optional) quality measures associated with different parts of the algorithm accompanied with signals that respective parts of the algorithm are subject to automatic improvement according to specified quality measure.

User model is defined for a particular application, so in the proposed approach there is no concept of an overall user model with all his/her characteristics. Instead, there is a set of application specific profiles expressed with a help of OWL ontologies. There is an interesting possibility to (partially) fill user model for a new application, based on user model from the applications the user already took part, it can potentially be done in the phase of application deployment and crowd forming, but this is a subject to further research.

Task model describes task attributes and task structure, if it is relevant. In geospatial context, task model usually contains attributes holding physical location of the place task refers to.

The whole computation workflow is defined in a declarative way, using attributes defined by the task and user ontologies.

The framework already contains a number of typical workflow patterns that can be easily reused either in their default form, or specialized in a way, supported by the pattern. For example, pattern *Divide-And-Conquer* should be specialized by providing specific procedures for task decomposition and subtask answers composition.

6 Examples of Geoinformation Tasks

The proposed framework can be used for geospatial information processing. In this section, it is shown how to build a crowd sensing application with the help of this framework.

The application described in this section used mostly for the demonstration purpose. Its main feature is to address a free-form request to people that are located in specific spatial area. An application like that can be used in variety of scenarios, for example, to query for a price (or availability) of some product in a shop, that doesn't expose its product listings to the internet, to find out traffic situation, etc. For the sake of simplicity, the expected answer should be in the form of one real number.

As it was discussed in Sect. 5, an application for the proposed framework consists of a user model, a task model and a workflow code (the forth component is optional, and is not covered in this example). The user model includes all the user features relevant for the application. In this particular case, no special competency modeling should be done, the only user property that is relevant is the user's ability to understand written language of the free-form question. Therefore, the user model in this case consists of the set of languages, which the user is able to understand, and the user's geographical coordinates.

The task model for this particular application is also simplistic, it is represented by three attributes—the text of the question, its language and the spatial area the request refers to.

To organize the application workflow it is convenient to use the *redundancy-based quality control* pattern. This pattern requires specialization by the redundancy number and the procedure of answers reconciliation. For example, it is possible to set the redundancy number to three and the reconciliation procedure as finding a mean value of answers, provided in redundancy branches of execution (remember, that answers for this question should be real values). Application workflow also specifies condition of crowd member selection for tasks, which is crowd member coordinates must be inside spatial area the question refers to and crowd member's list of languages must contain the language of the question.

Crowd members must join to this application to avail themselves as request answerers. When the problem setter wants to send a request, he/she provides input to the application, that is request text, its language, and reward. All this input values form a task instance. This instance is then processed by program interpreter and transformed into three human tasks. The task is sent through the crowd platform to crowd members satisfying the selection conditions, the answers are received, averaged according to workflow definition and presented to the problem setter (Fig. 2).

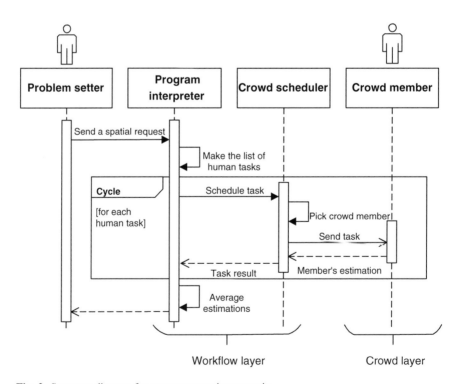

Fig. 2 Sequence diagram for request answering scenario

The provided example illustrates design of a simplified crowd computing application, but the flexibility of the proposed framework allows to define more complex workflows and models.

For instance, the provided example can be easily transformed into a crowd-sourced generation of a thematic map (relating to a free-form request). For this purpose, user model and task model remain unchanged, but the workflow description includes the step of task separation (*Divide-and-conquer* pattern) according to some level of spatial discretization. After that separation a bunch of subtask with the same structure but smaller spatial extent is generated. So, the area, that was provided by the problem-setter, is split into smaller areas. For each of this subtasks a *redundancy-based quality control* pattern is applied as in the original example. Resulting dataset will contain a set of human "measurements" that can be overlaid onto a geographic map forming a thematic layer related to the nature of the question explored by the problem-setter.

Another practical example of crowd computing with geospatial information can be assignment of physical area of competence to each member (e.g., ethnography, economics experts) and addressing questions to experts that have knowledge of particular area.

7 Conclusions

In the paper, a problem of crowd computing process organization was discussed with an application to geospatial information processing. The analysis of current developments revealed the set of specific issues that must be addressed in any crowd computing framework design and the set of crowd workflow patterns that are tried-and-tested ways to effectively address these issues. A set of universal requirements for a crowd programming environment was provided to generalize the work that has been performed so far in the area. Then, the original crowd computing framework design was described.

Although most of the stated requirements are addressed by the proposed crowd computing framework, the model to account for rewarding and incentivization still needs to be clarified and developed, especially for the case, when problem-setter and underlying crowd platform are using different rewarding mechanisms.

Another direction for further work is experimentation with different approaches and algorithms of self-organization and self-adjustment of the geoinformation crowd computing workflow.

Acknowledgments The research was supported partly by projects funded by grants # 13-07-00271, # 14-07-00345 of the Russian Foundation for Basic Research, and by Government of Russian Federation, Grant 074-U01.

References

1. Ahmad S, Battle A, Malkani Z, Kamvar S (2011) The jabberwocky programming environment for structured social computing. In: Proceedings 24th annual ACM symposium user interface software and technology, UIST '11 (ACM, New York, NY, USA, 2011), pp 53–64
2. Barowy DW, Curtsinger C, Berger ED, McGregor A (2012) AutoMan: a platform for integrating human-based and digital computation. ACM Sigplan Not 47(10):639–654
3. Bernstein A, Klein M, Malone TW (2012) Programming the global brain. Commun ACM 55 (5):41–43
4. Dean J, Ghemawat S (2004) MapReduce: simplified data processing on large clusters. In: 6th symposium on operation system design and implementation, San Francisco, CA, 2004
5. Faulkner M et al (2014) Community sense and response systems: your phone as quake detector. Commun ACM 57(7):66–75
6. Franklin M, Kossmann D, Kraska T, et al (2011) CrowdDB: answering queries with crowdsourcing. In: Proceedings of ACM SIGMOD conference
7. Gamma E, Helm R, Johnson R, Vlissides J (1995) Design patterns: elements of reusable object- oriented software. Addison-Wesley Longman Publishing Co., Inc., Boston
8. Gomes C, Schneider D, Moraes K, de Souza J (2012) Crowdsourcing for music: survey and taxonomy. In: Proceedings of 2012 IEEE international conference on systems, man, and cybernetics, Seoul, pp 832–839, 14–17 Oct 2012
9. Haklay ME, Antoniou V, Basiouka S, Soden R, Mooney P (2014) Crowdsourced geographic information use in government. Global Facility for Disaster Reduction and Recovery (GFDRR), World Bank, London
10. Kittur A, Smus B, Khamkar S, Kraut RE (2011) Crowdforge: crowdsourcing complex work, In: Proceedings 24th annual ACM symposium user interface software and technology, UIST'11 (ACM, New York, NY, USA, 2011), pp 43–52
11. Kucherbaev P, Tranquillini S, Daniel F, Casati F, Marchese M, Brambilla M, Fraternali P (2013) Business processes for the crowd computer. In: Business process management workshops, pp 256–267
12. Kulkarni AP, Can M, Hartmann B (2011) Turkomatic: automatic recursive task and workflow design for mechanical turk. In: CHI '11 extended abstracts on human factors in computing systems (CHI EA '11). ACM, New York, NY, USA, pp 2053–2058
13. Little G, Chilton LB, Goldman M, Miller RC (2010a) TurKit: human computation algorithms on mechanical turk. In: Proceedings of UIST (2010), ACM, New York, pp 57–66
14. Little G, Chilton LB, Goldman M, Miller RC (2010b) Exploring iterative and parallel human computation processes. In: Proceedings of the ACM SIGKDD workshop on human computation (HCOMP '10). ACM, New York, NY, USA, pp 68–76
15. Little G, Chilton LB, Goldman M, Miller RC (2009) TurKit: tools for iterative tasks on mechanical turk. In: Proceedings ACM SIGKDD workshop on human computation, HCOMP'09 (ACM, New York, NY, USA, 2009), pp 29–30
16. Marcus A, Wu E, Karger DR, Madden SR, Miller RC (2011) Crowdsourced databases: query processing with people. CIDR '11
17. Mavridis N, Bourlai T and Ognibene D (2012) The human-robot cloud: situated collective intelligence on demand. In: 2012 IEEE international conference on cyber technology in automation, control, and intelligent systems (CYBER), pp 360–365
18. Meier P (2012) How crisis mapping saved lives in Haiti. http://voices.nationalgeographic.com/2012/07/02/crisis-mapping-haiti/. Accessed 28 Nov 2014
19. Minder P, Bernstein A (2012a) How to translate a book within an hour: towards general purpose programmable human computers with CrowdLang, Proceedings of the 3rd Annual ACM Web Science Conference, Evanston, Illinois, pp 209–212, 22–24 June 2012
20. Minder P, Bernstein A (2012b) CrowdLang: a programming language for the systematic exploration of human computation systems. In: Proceedings of the 4th international conference on social informatics, Lausanne, Switzerland, pp 124–137, 05–07 Dec 2012

21. Parshotam K (2013) Crowd computing: a literature review and definition. In: Machanick P, Tsietsi M (eds) Proceedings of the south african institute for computer scientists and information technologists conference (SAICSIT '13), ACM, New York, NY, USA, 2013, pp 121–130
22. Scekic O, Truong H-L, Dustdar S (2013) Incentives and rewarding in social computing. Commun ACM, 56(6):72–82
23. Sengupta B, Jain A, Bhattacharya K, Truong H-L, Dustdar S (2013) Collective problem solving using social compute units. Int J Coop Inf Syst 22(4) (World Scientific Publishing)
24. Smirnov A, Levashova T, Shilov N (2015) Patterns for context-based knowledge fusion in decision support systems. Inf Fusion 21:114–129
25. Smirnov A, Ponomarev A, Shilov N (2014) Hybrid crowd-based decision support in business processes: the approach and reference model. Procedia Technol 16:376–385
26. Ushahidi (2014) http://www.ushahidi.com/. Accessed 28 Nov 2014
27. Zhang H (2012) Computational environment design. Ph.D thesis, Harvard University, Graduate School of Arts and Sciences, Cambridge, Massachusetts
28. Zhang Y, McBroom M, Grogan J, Blackwell PR (2013) Crowdsourcing with ArcGIS online for natural resources management. http://www.faculty.sfasu.edu/zhangy2/download/crowdsourcing.pdf. Accessed 28 Nov 2014

Dynamic Resources Management in Agile IGIS

Nataly Zhukova and Alexander Vodyaho

Abstract The paper is focused on the problem of resources management in environment of the intelligent geo information systems (IGIS) for solving complicated operational tasks in conditions of limited resources. Support of agile features in IGIS requires considerable amount of additional resources for building, modifying and executing context sensitive processes in dynamics. The proposed solution assumes management of the system processes using a set of business rules and policies that allow make comprehensive estimations of processes priorities based on their complexity and importance; reduce amount of calculations using revealed information and knowledge from historical data; distribute loading of the technical means due to preliminary data processing and analyses made according to expected behavior of end users. The dynamic resource management is implemented using means and tools of artificial intelligence that are a part of IGIS and ready program components of the middleware level provided by Globus Toolkit.

Keywords Agile GIS · Resource management · Measurements processing

1 Introduction

Modern state of the art of information systems (IS) is characterized by permanent expansion of the scope of geo information systems (GIS) usage. The systems are highly appreciated by end users as they support a convenient environment for their operation. Permanent expansion of the sphere of GIS usage, increase of complexity of the tasks to be solved, increase of functional and non-functional requirements imposed to the systems have led to appearance of a new class of IS—intelligent geo information systems (IGIS) [1, 2]. IGIS are knowledge based systems that contain a

N. Zhukova (✉) · A. Vodyaho
St. Petersburg Electro Technical University 5, Popova Str, St. Petersburg, Russia197376
e-mail: nazhukova@mail.ru

A. Vodyaho
e-mail: aivodyaho@mail.ru

© Springer International Publishing Switzerland 2015
V. Popovich et al. (eds.), *Information Fusion and Geographic Information Systems (IF&GIS' 2015)*, Lecture Notes in Geoinformation and Cartography,
DOI 10.1007/978-3-319-16667-4_8

wide range of means and tools of artificial intelligence. Modern IGIS are capable to support decision making, solve the tasks of situation recognition, classification, assessment and awareness, they can solve a wide range of applied mathematical problems, including radio location and hydro acoustics problems. To provide the enumerated services to end users many internal services that implement mathematical and imitation models, mathematical and empirical methods and algorithms, support interaction with external systems, represent results in 2D and 3D formats and a number of other specialized services [1, 3] such as logic inference machine are required. Traditionally solutions of the problems formed in IGIS are based on usage actual and historical data, cartographic and hydro meteorological information, information and knowledge of subject domains. Commonly IGIS include a set of business services to process data, information and knowledge [4] and a set of infrastructure services, which are used for tuning algorithms and business processes to context of the solved problems.

The list of the enumerated services shows the high level of complexity of IGIS though the list contains only a part of services implemented in real systems. The total number of services in IGIS can vary from several dozens to several hundreds. The majority of the services require considerable computational resources.

Recently an agile concept[1] was implemented in IGIS. Agile means "smart and clever; having a quick resourceful and adaptable character".[2] Four points of agility are considered for IGIS: application of agile methods and algorithms, support of agile business logic, agile systems architecture and agile approach to systems development [5, 6]. Support of agility features requires additional computational resources, because it is necessary to support complementary models and execute specialized algorithms to estimate system state; several iterations can be required to solve end users problems. For effective operation of IGIS dozen and hundreds of processes that refer to the tasks defined by different users are to be executed simultaneously. It is not a trivial task to manage processes in IGIS, because processes can have different priorities which depend upon context. The most part of the native schedulers refer to the level of operating systems (OS), for example, the schedulers of Linux (UNIX) systems. They are not able to support resource scheduling sensitive to the currently observed conditions that is required by IGIS. In IGIS it is necessary to use sophisticated knowledge based algorithms for resources management that allow distribute and continuously manage the resources at the level of tasks or business processes.

In the paper an approach to resources management of modern and perspective IGIS is suggested. The approach is based on application of extended capabilities of artificial intelligence supported in IGIS and a set of technological and program solutions recently developed in the sphere of IT. Due to usage of ready solutions the development and the implementation of the suggested approach doesn't require additional financial resources.

---

[1]http://agilemanifesto.org/.
[2]http://www.merriam-webster.com/dictionary/.

2 Common Approaches to Resources Management

Resources include all types of available technical means and information. Technical means that provide computational resources are commonly organized in the form of clusters or grids. Resource management provides efficient distribution of resources among processes created to solve end users problems. The main tasks of the resource management are planning the resources provisioning and tracking actual states of the resources. Three approaches to resource management are now commonly used in IS.

(a) Resource management based on using capabilities of the OS.
(b) Resources management based on using middleware developed by third-party organizations.
(c) Development of specialized new solutions.

OS provide program interfaces that give information about the processes and allow manage processes states and course of performance. For example, in Linux OS commands "top" and "ps" can be used to get the list of processes and detailed information about each of the processes. All processes in Linux have priorities that can be dynamically changed using the "nice[-adnice] command [args]" command. Processes can also be defined as background or active processes and can be terminated.

To support cluster and grid technologies specialized middleware can be used. It plays the role of a mediator between the subtasks that are executed on a set of remote computers. Middleware provides a number of program tools for exchanging messages in networks, call of remote procedures, management access to resources and other utilities. The leaders of the software for distributed computing are Globus Toolkit,[3] UNICORE[4] and gLite.[5] They all allow solve successfully the task of resources management, but they have different implementation. Globus Toolkit is a tool for development grid-based applications using service-oriented approach. UNICORE is focused on provisioning uniform access to computers. gLite provides a basis for distributed systems development. gLite is mostly used for scientific purposes.

Specialized middleware solutions are mostly developed by large-scale enterprises to meet all internal requirements. Specialized systems have been developed, for example, by SAS Institute Inc.[6] that is now one of the leaders in business analytics sphere. SAS has developed a comprehensive grid solution called "SAS Grid Computing" that runs on the grid nodes on Red Hat Enterprise Linux and Intel Xeon processor 5500 series. Experimental researchers made by SAS showed that

---

[3]http://toolkit.globus.org/toolkit/.
[4]https://www.unicore.eu/.
[5]http://grid-deployment.web.cern.ch/grid-deployment/glite-web/.
[6]http://www.sas.com/en_us/home.html.

SAS Grid Computing enables automatically leverage a centrally managed grid infrastructure and provides high availability for critical services and business processes.

Usage of capabilities provided by OS is sufficient to develop solutions for management separate servers. Development of complex resources management systems that support grid computing is highly expensive and can be fulfilled only by large-scale enterprises and corporations. In addition, effectiveness of the specialized solutions strongly depends on the OS and the hardware. Application of external program tools is cost effective but rarely meets all the requirements to resources management systems.

3 Resources Management in IGIS

IGIS systems are designed, developed and supported using agile approach. It allows create systems that have different architecture and can be deployed on one or several servers in one network, on multiple servers in a number of networks or can be executed using virtual technologies, for example, cloud technologies.

The main peculiar features of the IGIS from the point of view of resource management are the following:

(a) types and the structure of business processes and the timeline of their execution can be hardly defined a priori due to variety of the solved problems and wide sphere of IGIS application;
(b) the tasks solved in IGIS, in particular, decision making support, operating monitoring have high level of complexity. They require application of computational and imitation models, mathematical and empirical methods and algorithms, means and tools of artificial intelligence;
(c) the end users of IGIS have different levels of qualification and different skills in the sphere of IT. The majority of the users are not ready to consider the loading of the technical means and wait for the desired solutions;
(d) the backend technical resources used for IGIS deployment are commonly heterogeneous. The set of the used frontend technical means are defined by end users and can be of any type.

The main features of IGIS define the following requirements to the resources management:

(a) static adjustment of the systems to the available resources at the design stage;
(b) dynamic resources scheduling at the stage of system operation;
(c) dynamic adaptation of the used models, methods and algorithms to amount of available resources;
(d) possibility to extend and to modify logic used for resources management.

Resource scheduling is the key mechanism of resource management that must allow:

(a) define the priorities of the processes that relate to one task or a group of tasks;
(b) define the policies or business rules for resource management;
(c) estimate amount of resources required to solve the task with the defined quality indicators.

The problem of the resource management in IGIS can be considerably simplified due to:

(a) usage of information and knowledge about the solved problems and the subject domains provided by experts or mined from historical data. Data Mining techniques[7] are commonly used to reveal knowledge from historical data;
(b) loading of the technical means can be distributed in time using information and knowledge about the end users expected behavior. Behavior of the users can be described manually or on the base of the results of the analyses of the information about user's sessions.

For example, logic rules are used for solving tasks of protection of ships and vessels sailing under constrained conditions and in navigation dangerous areas; solving tasks of tactical maneuvering. Alternative ways to solve the tasks assume performance of large scale numerical calculations. Thus, vessels sailing near ports and in port entrances at small depth and low speed requires precise estimation of the influence of surface winds, underwater currents, sea bottom contour.

One of the most complicated tasks from the computational point of view is the task of atmosphere and ocean actual parameters estimation. Separate observations of atmospheric pressure, wind speed, temperature and salinity of water are used to generate horizontal gridded fields. The fields for the areas that are expected to be demanded by the users are calculated and/or improved as soon as operative measurements are received and free computational means are available. Most often operators request gridded fields for responsibility zones.

Resource management in IGIS can be implemented using approach based on capabilities provided by OS or a new service adapted to needs of IGIS can be developed.

Application of the first approach assumes provision of maximum possible resources to all processes. Due to diversity of the processes and technical means it leads to irrational usage of the resources and in separate cases does not allow solve the problems with the desired quality.

The second approach is more preferable, but it is more complicated. It assumes implementation of an interface for interaction with OS, means and tools for business rules and policies management.

---

[7]http://www.kdnuggets.com/.

In order to reduce resources required for design, development and support, the new service can be developed using ready middleware solutions, in particular Globus Toolkit, and the internal capabilities of IGIS. Globus Toolkit can provide interaction with OS. Means and tools of artificial intelligence (AI) provided by IGIS allow efficiently support resources scheduling.

4 Resources Management System Architecture

Agile IGIS for end users are designed and developed using an IGIS framework [6]. The framework allows create knowledge based architectural descriptions and implement end systems using ready technological and program solutions. The resource management system is considered to be a part of the framework.

Architecture of the resource management system is shown in Fig. 1. Its key components are: user stations, IGIS components for resource management, including resource manager, scheduler and dispatcher, middleware services, in particular, Globus Toolkit services, OS and hardware.

Operators can set tasks to the system and monitor system state. Experienced operators have access to information about available resources, their state and

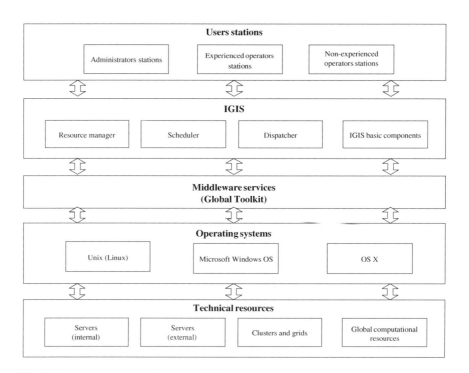

Fig. 1 Resources management system architecture

performance. Administrators have rights to schedule all tasks. It is possible to run multiple instances of one client at different locations.

Resource manager is the central component from where the resources are managed. It is responsible for creating and maintaining processes, interacting with clients, the scheduler and the dispatcher. The engine of the manager ensures that the states of the processes are recorded in persistent storage and restores the states in case the system goes down. Priorities of the process are calculated by the scheduler. The scheduler discovers and selects resources, assigns resources to processes. The dispatcher initiates the execution of the processes on the selected resources according to the defined priorities.

IGIS basic components include modeling and mathematical models services, inference machine [3]. The components are used by the scheduler to determine and update the priorities of the processes according to the changes of the contexts of the tasks execution.

Access to data, information and knowledge in IGIS are assigned to the services of data bases, knowledge bases and ontologies. The dynamic information model [2, 4] of the whole system that is sustained by the objects' server provides subject domain universal description. It supports inheritance mechanism, separation of the object's stationary and transient data, universal mechanism of relations and gives access to history of the objects properties' states and etc.

Globus Toolkit that forms the base of the middleware services is an open source toolkit for grid computing developed and provided by the Globus Alliance. It contains a set of tools for constructing grids, covering security measures, communications, resources location, resources management and etc. The protocols and functions defined by the Globus Toolkit are similar to those in networking and storage, but have been optimized for grid-specific deployments. Different OS, including Linux, Windows, OS X are supported by the toolkit.

5 Resources Scheduling

Logic of resource scheduling is based on the attempts to find multiple ways for solving each of the problems and select the best solutions according to the currently observed conditions. The state diagram of the resources scheduler is given in Fig. 2.

The tasks and their initial descriptions are passed to the scheduler by the resource manager. The scheduler gathers additional information required for processes building and executing, including all types of requirements and restrictions. Information is mostly provided by IGIS components. Using the detailed description of the tasks one or several alternative processes are preliminary defined. To build processes in dynamics agile techniques proposed in [7] are used. The parameters of the processes and the expected results are estimated. About each process a decision about its compliance with the requirements of the solved tasks is made. None, a

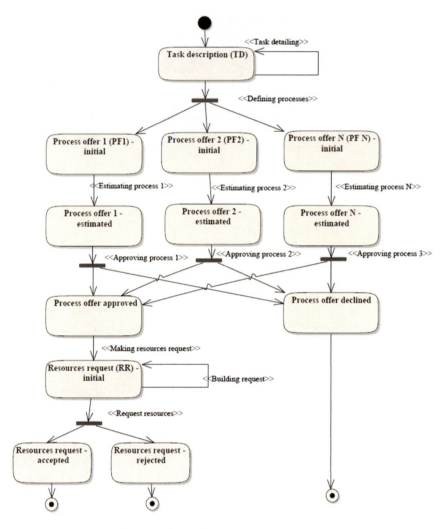

Fig. 2 State diagram of the resources scheduler

single or a set of processes can be selected from the set of preliminary processes. It's admissible to modify processes and estimate them iteratively. For the selected processes their priorities are defined. The algorithms for priorities estimation are considered below. The description of the resources required for processes execution are represented in a form of a request for the resources. In case availability of the resources is approved by the resource manager, the description of the processes is passed to the manager for further execution.

6 Algorithms for Priorities Estimation

Priorities of the tasks and the corresponding processes are defined for all new tasks at the moment of their creation. Different processes defined for the same task may have different priorities. Due to constantly changing conditions of the processes execution priorities of the processes have to be continuously updated.

The key factors that influence priority of a process are the following:

- estimation of process priority given by the owner of the process;
- priority of the group of users to which the owner of the process refers;
- conditions in which the solved problem was stated;
- capabilities of the end users to aware the results;
- conditions in which the formed results will be used;
- period of time during which the task is actual and after which the formed results are useless.

Algorithm for priorities estimation using information about priorities of the tasks and priorities of the tasks owners (Algorithm 1). The algorithm includes three steps:

(a) preliminary priorities of all tasks that can be solved by the system and the priorities of the groups of users are defined;
(b) at the start of the system work the queue of the tasks is formed. The tasks in the queue are ranged according to the preliminary defined priorities;
(c) during system usage when new tasks appear, the queue is rearranged taking into account the priorities of the new tasks and the priorities of the users that have set the tasks.

The priorities of the tasks and the groups of users can be defined in a table or represented in a form of functional dependencies. The table view of the priorities is given in Table 1.

In the table following notation is used: $T = \{T_i\}_{i=1}^{m}$—the set of the tasks that can be solved by the system; $K = \{K_i\}_{i=1}^{n}$—the set of the groups of users; m—total number of the tasks; n—total number of the groups of users; $Q = \{Q_{ij}\}$, $i = 1\ldots m$, $j = 1\ldots n$—the priorities of the tasks for different groups of users. Description of the Algorithm 1 is given in Fig. 3.

Table 1 Table view of priorities of the tasks and the groups of users

Tasks	Groups of users							
	K_1	K_2	K_3	...	K_i	...	K_{n-1}	K_n
T_1	Q_{11}	Q_{12}	Q_{13}	...		...	...	Q_{n1}
T_2	Q_{21}					...	...	
T_3						...	...	
...								
T_k					Q_{ki}	...	...	
...								
T_m						...	...	Q_{mn}

Input: *the set of tasks T, groups of users K, priorities of tasks for groups of users Q.*
Output: *the queue of the tasks that have to be solved E $E, E, ..., E_i, ..., E$ E T, K, Q,*
i 1..m, j 1..n, m_i - the total number of tasks. $_1$ $_2$ $_{m_1}$ $_i$ $_i$ $_j$ $_{ij}$
Step 1 *(defining priorities for the tasks and the groups of users)*
   set *priorities for the tasks T $\{T\}_{i=1}^{m}$ using information provided by the dynamic information model;*
   set *priorities for the groups of users K $\{K_i\}_{i=1}^{n}$ using information provided by the dynamic information model;*
   set *priorities of the tasks for the groups of users Q $\{Q_{ij}\}$, i 1..m, j 1..n;*
Step 2 *(ranging the tasks that have to be solved)*
   define *the set of the tasks T' that have to be solved, $T' : T'$ $\{T\}_{i=0}^{m_1}$, T' T, T', m_1 - the number of the tasks to be solved;*
   define *a set E' ;*
   for each T T', i 1..m$_1$
      define *the groups of users K_j K j 1..n ;*
         define *the priority Q_{ij} for the corresponding task T_i and group of users K_j ;*
         add *the element E_i' T, K, Q_{ij} $E' : E'$ E'_i ;*
                                 to the set
   range *the set E' ;*
   set E E' .
Step 3 *(modifying and updating the queue according to the new tasks)*
   define *the set of new tasks T'' that have to be solved, $T'' : T''$ $\{T\}_{i=0}^{m_2}$, T'' T, m_2 - the number of the new tasks to be solved;*
   define *the set E' ;*
   for each T_i T'', i 1..m$_2$
      define *the groups of users K_j K j 1..n ;*
         define *the priority Q_{ij} for the corresponding task T_i and group of users K_j ;*
         add *the element E_i' T, K, Q_{ij} $E' : E'$ E'_i ;*
                                 to the set
   add *existing tasks to the set $E' : E'$ E ;*
   range *the set E' ;*
   set E E' .

Fig. 3 Algorithm for priorities estimation based on information about priorities of the tasks and their owners (Algorithm 1)

Algorithm for priorities estimation based on analyses of the factors that influence the priorities and the factors weights (Algorithm 2). According to the considered list of factors that influence priorities a set of groups of priorities indicators can be defined (Table 2).

The approach to the priorities estimations based on analyses of the influence factors can be efficiently used, for example, to build dynamic map of sea situation. The map represents time, tactical and technical characteristics of ships and vessels and can be used to identify the potential dangers for the Naval Base responsibility zones. The main information sources are: radar stations, automatic identification

Table 2 Groups of priorities indicators

Group of indicators	Indicator	Indicator description	Indicator weight
The group that reflects the features of the tasks	I_1^1	A priori defined priority of the solved task	V_1^1
	I_2^1	Difference between required time and estimated time of completion	V_2^1
	I_3^1	Amount of the mismatches and their severity between the estimations of the solutions characteristics and requirements to the solutions	V_3^1
	I_4^1	Conditions in which the user has set the task; the indicator can consider different conditions depending on the applied subject domain	V_4^1
	I_5^1	Level of decomposition of the task to the subtasks and the process formed to solve the task to a set of sub processes	V_5^1
The group that reflects the features of the users	I_1^2	A priori defined priority of the group of users to which the owner of the task belongs	V_1^2
	I_2^2	Number of the consumers of the formed results not including the owner of the task	V_2^2
	I_3^2	The ratio of the number of the consumers of the formed results that are able/are not able to solve their tasks without the solution of the considered task	V_3^2
The group that reflects the features of the alternative solutions of the task	I_1^3	Number of the alternative solutions of the task	V_1^3
	I_2^3	Amount of the additional resources required by alternative solutions	V_2^3
	I_3^3	The minimal amount of resources required to solve the task	V_3^3
	I_4^3	Amount of information required to solve the task that can be provided by the dynamic information model	V_4^3
	I_5^3	Probability of successful gathering all information that is necessary to solve the task	V_5^3
	I_6^3	Cost of gathering information that is necessary to solve the task	V_6^3
The group that reflects risks that can realize at the stage of the processes execution	I_1^4	Possibility of change of the amount of required resources at the stage of the processes execution	V_1^4
	I_2^4	Cost of processes suspending and recovery	V_2^4

systems, external interacting systems, information portals such as Marine Traffic[8] and Ship Finder.[9] The total number of ships and vessels is more than half a million. Each of them are characterized with a set of parameters including coordinates,

---

[8]http://www.marinetraffic.com.
[9]http://www.shipfinder.com/.

trajectory, destination, name, type, status, speed, course. The information about the vessels is continuously updated.

It's obvious that the characteristics and behavior of all ships and vessels can't be analyzed. The priority of the vessels are defined according to

i. their distance to the responsibility zone and maximal speed of the vessels taking into account strength of wind and wind direction;
ii. amount of available information. The vessels that are detected by radar stations and are not identified have the highest priority;
iii. the type of the vessels. A vessel can be of the type of a passenger vessel, cargo vessel, tanker, high speed craft, tug, piloted vessel, yacht, fishing vessel. Operators have to pay special attention to high speed crafts and yachts.

Indicator I_1^1 is defined according to the priority of the vessels. I_2^1 depends on the total number of vessels and the time period during which the results of the analyses must be updated. The severity of the mismatches between the characteristics of the solutions and the requirements to the solutions (I_3^1) depend on the location and technical characteristics of the vessels. Indicator I_4^1 is set taking into account:

- the correspondence of the task to the sphere of the operator responsibilities;
- the operational situation in the region of the ships and vessels location;
- the position and the rank of the operator.

The level of the decomposition of the task of the vessels characteristics analyses I_5^1 is equal to all vessels.

Description of the Algorithm 2 for priorities estimation using indicators and their weights is given in Fig. 4. Priorities can be defined using (i) functions for priorities calculation; (ii) iterative ranging of the tasks according to the indicators weights and values based on application of radix sort algorithm [8].

Algorithm for dynamic estimation of priorities (Algorithm 3). The algorithm is based on finding solution that is close to the optimal solution in the conditions of defined restrictions. The target function identifies the set of tasks that can be solved without violating requirements to the formed results.

To formalize the problem let's consider n users. Each user sets a single task. s resources are available to solve the tasks. The states of the solved tasks at the time t can be described with the vector $\bar{x}(t) = \mathrm{col}(x_v(t), v = \overline{1,n})$, n—the number of tasks set by the users. Each component of the vector $x_v(t)$, $v = \overline{1,n}$ defines the state of the task of the v-th user. Let's consider a matrix of the controlling parameters that has $n \times s$ dimension $U(t) = (u_{v\mu}(t) : u_{v\mu} \in \{0;1\}, v = \overline{1,n}, \mu = \overline{1,s}) = [u_v, v = \overline{1,n}] = [u_\mu, \mu = \overline{1,s}]^T$ Solution of the problem can be described using non-stationary vector-matrix linear Cauchy confluent equation. The equation is non-stationary vector-matrix linear equation: $\bar{x}(t) = B(t) \circ U(t)$, $B(t) = (b_{v\mu}(t) : b_{v\mu} = \varepsilon_{v\mu}(t) \cdot \varpi_{v\mu}, v = \overline{1,n}, \mu = \overline{1,s})$—$n \times s$ dimension control matrix that defines the capability and the intensity $\varpi_{v\mu}$ of solving the v-th task using the s-th resource in the conditions $E(t) = (\varepsilon_{v\mu}(t) : \varepsilon_{v\mu} \in \{0;1\}, v = \overline{1,n}, \mu = \overline{1,s})$ considered for the defined time interval.

Input: the set of the tasks T, indicators used to calculate priorities $I\{I_i\}$, weights of the indicators $V\{V_i^j\}$, $f = 1,...,4$, $j = 1,...,G_f$, $|G_f|$ - a number of indicators in the group f.
Output: the queue of the tasks that have to be solved. $E = E_1, E_2, ..., E_l, ..., E_{m_1}$, $E_l = \langle T, K, Q \rangle$,

$i = 1..m$, $j = 1..n$, m_1 - the total number of tasks.

Step 1 (defining the set of indicators and the rate of their influence on the priorities)

1.1. **define** the set of the indicators that have to be considered for priorities estimation $I' \subseteq I$;

1.2 **define** a set V';

  1.3. **for each** $i_k \in I'$, $k = 1,...,s$, s - total number of the considered indicators;

    define the weight V_k, $V_k \in V'$; k

1.4 **normalize** the weights $V' = norm(V')$;

Step 2 (preparing data for priorities calculation)

2.1. **define** a matrix of indicators values U;

  . **define** the set of the tasks T' that have to be solved, $T': T' \subseteq \{T\}_{i=0}^{m_1}$, $T' \subseteq T$, T', m_1 - the number

of the tasks to be solved;

  for each $i_k \in I'$, $k = 1,...,s$

    for each $T_l \in T'$, $l = 1..m_1$

      define the value of the indicator u_{lk}, $U = u_{lk}$;

Fig. 4 Algorithm for priorities estimation using indicators and the indicators weights (Algorithm 2)

A component $b_{v\mu}(t)$ of the matrix $B(t)$ equals $\varpi_{v\mu}$ in case at defined time t the resource μ can be used to solve the v-th task with intensity $\varpi_{v\mu}$. The operation $[] \circ []$ is the operation of matrix multiplication: $B(t) \circ U(t) = \sum b_{v\mu}(t)u_{v\mu}(t)$, $v = \overline{1,n}$. The set of the considered restrictions contains: (i) technical, technological and informational restrictions imposed by the system; (ii) time restrictions defined by the time intervals during which the tasks are actual; (iii) edge conditions. Technical restrictions are restrictions of communication channels; restrictions caused by nonoperable state of technical means or usage the means for solving different tasks. Availability of data, information and knowledge required for solving the tasks defines the set of information restrictions. The restrictions determined by lack of methods, algorithms or procedures for solving the tasks refer to technological restrictions. Technical, technological and informational restrictions reduce the set of admissible alternative solutions: $U_\Delta = \left\{ u_{v\mu} : u_{v\mu} \in \{0,1\}; u_v^{-T} \cdot \bar{e}_{<s>} \leq 1; u_\mu^{-T} \cdot \bar{e}_{<n>} \leq 1 \right\}$. Boolean restrictions of the controlling parameters assume that one resource unit is able to solve a single task at one moment of time: $u_\mu^{-T} \cdot \bar{e}_{<;n>} \leq 1$, $\forall \mu = \overline{1,s}$. In case a resource can be used to solve more than one task simultaneously it is considered as a number of resource units. It is also supposed that at one time point a task can be solved only using one resource unit: $u_v^{-T} \cdot \bar{e}_{<s>} \leq 1, v = \overline{1,n}$. If execution of a task can be parallelized the tasks is considered as a number of tasks. The vector $\bar{u}_\mu = col(u_{v\mu}, v = \overline{1,n})$, $\mu = \overline{1,s}$ contains the elements of the matrix U (t) columns and the vector $\bar{u}_\mu = col(u_{v\mu}, v = \overline{1,s})$, $\mu = \overline{1,n}$—the elements of the transposed rows of the matrix. $\bar{e}_{<s>}$ and $\bar{e}_{<n>}$ are single vectors of the dimensions s and n correspondingly.

normalize *the indicators values* $U' = norm(U)$;

Step 3 (i) *(defining the priorities using functions for priorities calculation based on indicators and their weights)*

3.1. **define** E';

  for each $T_i \in T'$, $i = 1..m_1$, m_1 - *the number of the tasks to be solved*

    calculate the priority Q_i' *for the task* $T_i \in T'$ *using equation:* $Q_i = \sum_{k=1}^{s} u_{ik} \cdot V_k'$; *if the indicator is not supposed to be used, the weight of the corresponding indicator is considered equal to zero;*

    define *the groups of users* $K_j \in K$, $j = 1..n$;

      define *the priority* Q_{ij}' *for the corresponding task* T_i *and users group* K_j;

      add *the element* $E_i' = \langle T, K, Q_{ij}\rangle$ *to the set* $E' : E' = E' \cup ;_i$

  set $E = E'$.

Step 3 (ii) *(defining the priorities by iterative ranging of the tasks according to the indicators weights and values)*

3.1. **define** *an ordered set* E';

  sort *the indicators in the descending order* $I^* = sort(I')$ *according to their weights;*

  define $U' = reorder(U)$ *according to the order of* I^*;

  define $U'' = MSDRadixSort(U')$ *according to the order of* I^*, *MSDRadixSort* - *most significant digit radix sort algorithm;*

3.5. **for each** $T_i \in T'$, $i = 1..m_1$,

    define *the priority* $Q_i = m_1 - pos(T)$, $pos(T)$ - *the position of the task* T *in* U'';

    define *the groups of users* $K_j \in K$, $j = 1..n$;

      define *the priority* Q_{ij}' *for the corresponding task* T_i *and users group* K_j;

      add *the element* $E_i' = \langle T, K, Q_{ij}\rangle$ *to the set* $E' : E' = E' \cup ;_i$

3.6 **set** $E = E'$.

Step 4 *(modifying and updating the queue of the tasks)*

  define *the set of the new tasks* T'' *that have to be solved,* $T'' : T'' = \{T\}_{i=0}^{m_2}$, $T'' \subset T$, m_2 - *the number of the new tasks to be solved;*

    for each $i_k \in I'$, $k = 1,...,s$

      for each $T_i \in T''$, $i = 1..m_2$

        define *the value of the indicator* u_{ik}, $U = U \cup u_{;ik}$;

      $T' = T''$

      normalize *the indicators values* $U' = norm(U)$;

4.5 **define** *the list of tasks according to step 3*

Fig. 4 (continued)

The time restrictions limit the time periods during which the tasks are supposed to be actual for the users: $t \in T = [t_o, t_f]$, t_o and t_f are the beginning and the end time of the time periods.

Edge conditions define the possible changes of the resources state at the time of the task execution: $\bar{x}(t_0) = \bar{x}_0 \in X$, $\bar{x}(t_f) \in X$, $X = [\bar{x}_0, \bar{x}_f^d]$, $\bar{x}_f^d$—required complete solution.

The requirements to the set of the solved tasks can be described using Mayer function: $J = \left(\bar{x}(t_f) - \bar{x}_f^d\right)^T \cdot \Lambda \cdot \left(\bar{x}(t_f), \bar{x}_f^d\right)^T$. The function defines proximity of the formed solution $\bar{x}(t_f)$ to the required solution $\bar{x}_f^d \cdot \Lambda = diag(\alpha_v, v = \overline{1, n})$—the diagonal matrix of the a priori defined priorities of the tasks. The criteria for the selection of the optimal U_{opt} sequence of the tasks execution can be formulated as:

$$U_{opt} = \arg\min_{U \in U_\Lambda} J\left(\bar{x}(t_f), \bar{x}_f^d\right).$$

The optimization task can be solved using maximum principle of L. Pontryagin. It allows represent the control task defined on the continuous set of alternatives in the form of an edge task that can be solved with a limited number of iterations. The complexity of using the Pontryagin's principle and admissibility of nonoptimal solution makes it reasonable to use the branch and bound method [9].

To use the branch and bound method it is necessary to choose (i) the approach for estimating target functions for separate nodes; (ii) the approach for branching that defines the algorithms for choosing nodes for branching and branching parameters. In order to choose the nodes the nonlinear conjugate gradient method can be used. At the initial step of the method application the matrix of controlling parameters $U(\tau)^r$ at the time τ is considered. The index $r \in [1, 2, 3, \ldots]$ defines the number of the node.

The condition of the integer type of the values of the matrix elements is lifted. The elements of the vector of the controlling parameters are calculated using the equation: $u(\tau)_{v\mu}^{r(k)} = u(\tau)_{v\mu}^{r(k-1)} + \Delta u(\tau)^k$, $k = 1, 2, \ldots$. For each k-th element of the vector the value $u(\tau)_{v\mu}^{r(k)}$ is calculated. It is compared to the current record $u(\tau)_{v\mu}^{r(*)}$. In the case $u(\tau)_{v\mu}^{r(k)} > u(\tau)_{v\mu}^{r(*)}$ the value $u(\tau)_{v\mu}^{r(k)}$ is considered to be the record $u(\tau)_{v\mu}^{r(*)} = u(\tau)_{v\mu}^{r(k)}$.

The elements of the vector are iteratively calculated until the difference between the values obtained on neighboring iterations exceeds the predefined positive threshold: $u(\tau)_{v\mu}^k - u(\tau)_{v\mu}^{r(*)} \leq \delta u$, $\delta u \ll 1$. The threshold defines the precision of the calculations. For the parameter $\Delta u(\tau)^{(k)}$ that identifies the size of the step and the direction a fixed value is used: $\Delta u(\tau)^{(k)} < 1$. The direction of the gradient is calculated using the equation: $\Delta u(\tau)^{k-1} = \begin{cases} -\Delta u^{(k)}, \text{ in case } F\left(u(\tau)_{v\mu}^{r(k)}\right) \leq F\left(u(\tau)_{v\mu}^{r(*)}\right) \\ \Delta u^{(k)}, \text{ in case } F\left(u(\tau)_{v\mu}^{r(k)}\right) > F\left(u(\tau)_{v\mu}^{r(*)}\right) \end{cases}$, $F\left(u(\tau)_{v\mu}^{r(*)}\right)$—current record value of the target function.

To build branches one-sided branching method is applied [9]. The method assumes that the roots for branching are defined a priori. One-sided branching with the defined roots is based on consecutive fixing of parameters. Positive values are

Input: the set of the tasks T, the indicators used to calculate priorities $I_j^k \{I_j^k\}$ k 1,...,4,
j 1,..., $|G_j|$, $|G_j|$ - the number of indicators in the group G_j.
Output: the queue of the tasks that have to be solved. E $E_1, E_2, ..., E_l, ..., E_{m_1}$ E_l T_i, K_j, Q_{ij}

i 1..m, j 1..n, m_1 - the total number of tasks

Step 1 *(initializing step)*

1.1. **set** all elements of the matrix $U( )$ equal to zero;
1.2 **define** the initial record equal to minimal possible value or minus infinity $( )$;
1.3. **define** the set of the tasks T' that have to be solved, $T': T' \{T\}_{i\,i\,0}^m$, T' T_i, T', m_1 - the number

of the tasks to be solved;
1.4 **prepare** data according to Step 2 Algorithm 2;
Step 2 *(defining the queue of the tasks)*
2.1. **calculate** $U( )^*$ using the branch and bound method;

2.2 **define** E according to the calculated $U( )^*$;
Step 3 *(modifying and updating the tasks queue)*
4.1 **execute** the steps 4.1-4.4 of the algorithm 2;
4.2 **define** the list of tasks according to step 3

Fig. 5 Algorithm for dynamic estimation of priorities (Algorithm 3)

assigned to the parameters and the estimation task is solved. The method is selected as it requires minimum amount of calculations.

The description of the Algorithm 3 is given in Fig. 5. The algorithm doesn't allow take into account priorities of the groups of users. The tasks of each group of the users have to be analyzed separately in decreasing order of the groups' priorities.

The considered algorithms have different complexity and flexibility. The first algorithm does not require any additional resources. The third algorithm is flexible, but it requires resources for solving the optimization task.

7 Implementation of the Resources Management System

The middle layer of the resources management system contains GRAM (Globus Resource Allocation Manager), GSI (Grid Security Infrastructure), MDS (Monitoring and Discovery Service), GRIS (Grid Resource Information Service), GIIS (Grid Index Information Service) and GridFTP (Grid File Transfer Protocol) components provided by Globus Toolkit.

Knowledge based components of the logic layer, in particular, the resource manager and the scheduler use means and tools of AI. Information about the subject domain of resources management and information about the applied subject domains are stored in the system of IGIS ontologies. Rules used for resource scheduling form a separate part of the knowledge base. Logical inference is realized by the inference machine based on the Rete algorithm. IGIS modeling and mathematic models server executes algorithms for processes priorities estimation. All process for solving end users tasks are described in the form of scenarios [10]. Scenarios can be executed and managed using IGIS graphical tools. The dispatching of the processes according to the schedule is organized by the dispatcher component that interacts with IGIS management server.

Information about the processes, means for their monitoring, management, modification, analyses of the intermediate and the final results are available to the users through GIS interface.

Interconnections between the resources management system components and IGIS components at the backend are given in Fig. 6, where IGIS components are dashed.

The implementation of the scheduler supports a four step workflow (Fig. 7):

(a) estimating the complexity of the task, required and available resources;
(b) defining a set of alternative business processes for solving the tasks;
(c) estimating the priority of the processes;
(d) describing the required resources.

Estimation of the required and available resources is based on the computational complexity of the algorithms and planned utilization of the capacity of the technical means.

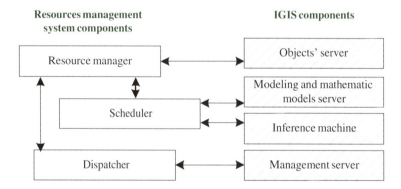

Fig. 6 Interconnections between the resource management system components and IGIS components

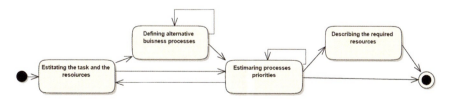

Fig. 7 Workflow of the scheduler

To estimate processes priorities in the scheduler three approaches are implemented. Approaches correspond to the considered algorithms for the priorities estimation.

(a) The priorities for separate tasks and/or groups of users are defined apriori in configuration files, for example, XML files. This approach is the simplest one and can be implemented using finite-state machine.
(b) The priorities are estimated by the system using a set of business rules or policies defined on the base of business rules.
(c) The business rules or policies are defined dynamically by the system using information and knowledge about the solved tasks and available resources provided by the dynamic information model.

To define alternative business processes for solving input tasks a set of basic rules is suggested (Table 3).

The set of the rules can be extended and the rules can be modified.

List of practical recommendations for building processes and defining their priorities include four main recommendations that are given in the Table 4.

Table 3 Business rules for building alternative processes

Rule №	Rule description
Rule 1	Precedents must be found and taken into account. If similar tasks in the same conditions have already been solved formed results can be used
Rule 2	Available knowledge can significantly help to solve the tasks. All knowledge that directly or indirectly refers to the tasks must be used
Rule 3	Possible ways to simplify the processes must be analyzed. In case of lack of the resources execution of some steps of the processes can be skipped or delayed
Rule 4	The steps of the processes that have been delayed must be executed as soon as possible
Rule 5	Some of the processes can be executed in advance. The behavior of the end users can be predicted and the expectable tasks can be solved

Table 4 Practical recommendations for building processes

Rec. №	Recommendation description
Rec 1	Ensure that the system doesn't have ready solutions that can be provided by the dynamic information model. The processes that have been executed to solve different tasks could acquire desired data or information. The solution could have been formed in advance in case the solved task is expectable
Rec 2	Ensure that that the external systems with which the system is able to communicate don't have ready solutions
Rec 3	Decompose the processes into a set of separate sub processes as it is much more convenient to execute structured processes
Rec 4	Avoid using complicated internal structures for data and information representation in the processes as it requires much resources for processes suspending and recovery

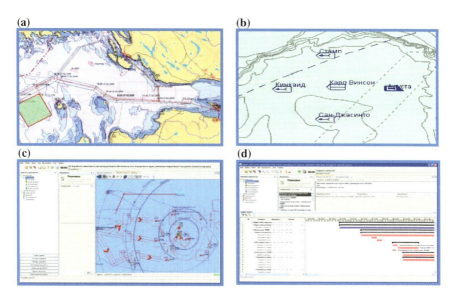

Fig. 8 GUI of Ontomap-V1

8 Case Study

On the base of IGIS a software complex for information support and automation of staff personnel functional activity at naval bases' has been developed (Ontomap-V1).[10] The complex is intended for staff personnel practical training aimed at making well substantiated decisions and planning operations of Naval forces at running the combat missions. A number of graphical forms implemented in Ontomap-V1 are given in Fig. 8. In the processes of operations planning a number of command centers are involved. The centers refer to operational-strategic, operational, operational- tactical and tactical levels.

Operation planning assumes solving tasks of sea situations data acquisition and processing, evaluation of the situations, operational computations and analysis of action variants, concepts formation and decision making, actions planning, cooperation organization and maintenance. To estimate the planed actions of naval forces imitational modeling subsystem is used.

To manage all processes required for solving tasks of the operations planning a resources management system is required. The queue of the tasks is organized according to

---

[10]http://oogis.ru/.

(i) operational information about surface, underwater, hydrometeorologic, mine, chemical, bacteriologic, and radiation situation provided by detecting facilities, maritime observation systems and interacting systems;
(ii) the timeline of the command centers activities;
(iii) dependencies of the command centers activities.

Sequential execution of the processes according to the time of their creation can lead to lack of the computational resources.

9 Conclusions

In the paper the approach to dynamic resources management is proposed. It allows improve performance of the computational resources, thus, reduce amount of required resources, fasten the decision making support processes and provide better solutions to IGIS users. The main advantages of the approach are the following:

- scalability. The scale of the system is defined by amount of available knowledge about resources scheduling. New information and knowledge can be added to the system using standard editors;
- low cost of the development and support. Ready solutions are used for implementation of the middleware. The logic level of the system is based on IGIS components. Only few new components are required;
- easy to use. The resource management system uses convenient IGIS environment to represent data and information to the users;
- high level of integration with IGIS. The system is implemented as a separate service and is included into the set of IGIS internal services. The service supports the protocol used for services interaction in IGIS;
- flexibility. Interaction with the components provided by external developers is organized using unified program interfaces and protocols. The set of the used components can be easily replaced by any other components that provide similar functionality.

The future work is oriented on expansion of the knowledge base and development of new algorithms for dynamic defining of the policies used to estimate the processes priorities.

References

1. Popovich V, Potapichev S, Sorokin R, Pankin A (2005) Intelligent GIS for monitoring systems development. In: 10th International conference on urban planning and regional development in the information society (Austria, Vienna, 22–25 Feb 2005). CORP, Vienna, pp 49–56
2. Popovich V (2014) Intelligent GIS conceptualization. In: Proceedings of the 6th international workshop on information fusion and geographic information systems: environmental and urban challenges. (SPb., 12–15 May 2013). Springer, Berlin, pp 16–44

3. Popovich V, Pankin A, Potapichev S, Galiano F, Zhukova N (2012) Service-oriented architecture of intelligent GIS. In: Proceedings of international symposium on service-oriented mapping (Austria, Vienna, 22–23 Nov 2012). JOBSTMedia, GmbH, pp 297–311
4. Popovich V, Voronin M (2005) Data harmonization, integration and fusion: three sources and three major components of geoinformation technologies. In: Proceedings of international workshop on information fusion and geographic information systems (Spb., 25–27 Sept 2005). Springer, Berlin, pp 41–46
5. Vodyaho A, Zhukova N (2014) Implementation of JDL model for multidimensional measurements processing in the environment of intelligent GIS. Int J Concept Struct Smart Appl 2(1):36–56
6. Vodyaho A, Zhukova N (2014) Building smart applications for smart cities—IGIS-based architectural framework. In: Proceedings of 19th international conference on urban planning and regional development in the information society (Austria, Vienna, 21–23 May 2014). CORP., Vienna, pp 109–118
7. Vitol A, Pankin A, Zhukova N (2014) Adaptive multidimensional measurements processing using IGIS technologies. In: Proceedings of the 6th international workshop on information fusion and geographic information systems: environmental and urban challenges (SPb., 12–15 May 2013). Springer, Berlin, pp 179–200
8. Preiss B (1999) Data structures and algorithms with object-oriented design patterns in C++. Wiley, Chichester, 688 pp
9. Clausen J (1999) Branch and bound algorithms—principles and examples. Technical report. University of Copenhagen, 30 pp
10. Sorokin R, Ivakin Y (2005) Application of artificial intelligence methods in geographic information systems. In: Proceedings of international workshop on information fusion and geographic information systems (SPb., 25–27 Sept 2005). Springer, Berlin, pp 105–114

Part IV
Applying GIS Technologies to Various Application Domains

Application of GIS Technologies in Historic and Ethnographic Research

Yan Ivakin and Vladislav Ivakin

Abstract Integration of the two greatly different directions in the modern science like history and geoinformation systems allows for developing and implementing a number of qualitatively new information systems applicable to the historical and ethnical research. The here presented paper is aimed at considering basic potential and perculiarities of the above information technologies.

Keywords Geographic information system · GIS technologies · Historic reconstruction · Earth remote sensing · GIS-based representation of knowledge · Data about historical · Geographic process

1 Introduction

Geographic information systems (GIS) have been widely applied in a capacity of the software tools accessible to a range of professionals; they also are a subject of a thorough attention of natural and humanitarian scientists The above attention to a certain degree has stimulated a rapid growth of the GIS technologies. Currently the software producers develop either their own GIS or a sort of intermediate software meant for a design of the systems that realize various innovative information technologies. Traditionally Geoinformation systems used to be developed as means aimed at a visual representation of the geographic (space coordinated) information. (General meaning content of the Geoinformation system assumes an information system that implements acquisition, storing, processing, access to, mapping and

---

Y. Ivakin (✉)
St. Petersburg Institute for Informatics and Automation
of Russian Academy of Sciences (SPIIRAS), St.Petersburg, Russia
e-mail: Ivakin@oogis.ru

V. Ivakin
Lomonosov Moscow State University (Department of History), Moscow, Russia
e-mail: Ivakin-11@mail.ru

dissemination of space-coordinated data [1, 5]) Most out of the well-known and widely used GISs are first and foremost the software tools that visualize the digital data sets of the geographic information. However, an analysis of the modern requirements to these systems that are considered as a research toolkit in the historical science, anthropology, and ethnography reveals a growing necessity to use GIS along with its conventional application to visualization and historic and geographic processes modeling as a platform integrating the heterogeneous information that underlies the above processes study. Just this possibility allows for considering GIS as a prognostic tool intended for a substantiation of research decisions, newly discovered facts, knowledge. Namely the properties generally perceived by a researcher as a basically new quality are necessary and specific for the geoinformation systems that allow for integrating data and knowledge about space processes of a disparate (including historic, anthropologic, ethnographic and other) nature.

Particular trend in the GIS technologies development assumes an implementation of the techniques and tools at the heterogeneous information integration and fusion intended for an extension of their functional scope. The here presented paper proposes to use in a capacity of geoinformation system oriented to the historic and ethnographic research area the totality composed of GIS-interface, system of space and time modeling, expert system and ontology editor. At that, the expert system plays a role of a structural component that assures a new quality at integration of the multi-faceted and heterogeneous information and includes [2, 7, 8]:

- software full functional inference machine; inference machines of modern artificial intelligent interactive environments like CLIPS, Jess and other can be used in a requested capacity;
- ordered collections of knowledge bases for various scenarios in the course of historic and geographic processes.

It should be noted that in the considered case the expert system is used as a conventional tool that intellectually supports the researcher as well as the modeling control system that operates in accordance with some "scenarios".

Use of the given architecture GIS aimed at information integration and fusion in a course of historic and ethnographic research resulted in an emergence of quite a number of new information technologies namely:

- Technology representing the Earth remote sensing data about the space extended historic objects;
- Technology for rendering (refinement) boundaries of the areas having ethnographic, political, economic and other values;
- Technology for modeling the retrospective dynamics of various historic as well as historic and geographic processes;
- Technology for identification and refinement of the facts in the development of historic and geographic processes based upon visualization of their textual descriptions;
- and other.

The above technologies of computer science are widely applied in various research areas such as history, ethnographic and Digital Humanities. Detailing of the composition and content of the indicated information technologies allows for describing the specifics of GIS methods' application as well as of the relevant information integration tools oriented to the use in the historic and ethnographic research sphere in whole.

2 Technology Representing the Earth Remote Sensing Data About the Space Extended Historic Objects

Technology representing the Earth remote sensing data (ERS) about the space extended historic objects supports the retrospective study of the state and dynamic changes in those objects and phenomena whose geographic dimensions allow for observing them in the large exclusively from the near-earth orbit. The following objects and phenomena can serve as examples, namely the Great Wall, artificial islands at the coast of the United Arab Emirates, changes of the objects' infrastructure within the areas of extensive natural cataclysms, etc.

The following sequence of stages is separated in an aggregate of the given technology:

Realizing an orientation of the source digital set for ERS snapshot within the GIS coordinate system and its binding to the electronic map with an allowance for the used Earth model (Krasovsky Ellipsoid (SK-42, SK-95), WGS-84 and other.) and cartographic projection (autogonal (Mercator), lateral and cylindrical (Gauss-Krueger), tapered and other). The essence of this stage's research is shown in Fig. 1.

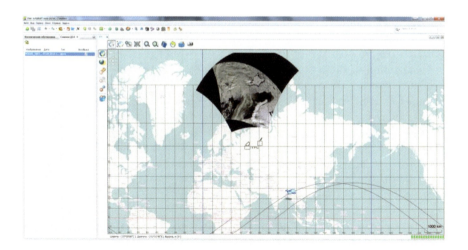

Fig. 1 Orientation and ERS snapshot binding in GIS

Fig. 2 Space snapshot of the Great Wall fragment prior to filtration

At that, each ERS snapshot is considered as a specialized set of space data that possess necessary attributes allowing for the positioning on the electronic map (Location and altitude of the satellite the snapshot is taken from, coordinates of the reference binding points, exact snapshot time, etc.). To describe such attributive data the specialized well-known text format (WKT) is used.

Separation of a digital set matching a definite area of the historic or ethnographic object under study. on the earth surface source snapshot (if resolution permits). Figure 2 depicts a snap of the Great Wall fragment that passes through a mountain terrain as an example of such a set of the source ERS data.

Tuning out (filtration) of the imagery for the object under study of spectral and atmospheric disturbances, visual and contrasting imagery refining. The idea of this stage is given in Fig. 3;

Recognition of the historic or ethnographic object under study and its semantically important elements on the process snap by the images' sequential segmentation, forming the features' set for the object separation (elements) and classification based on these features. As a rule this stage is the most complex one and labor consuming in ERS data representation on the electronic map. With due perfection of the modern technologies of pattern recognition and appropriate software readiness this stage assumes a direct human involvement at the final decision on attributing this or that image to the pattern of the being studied object (Fig. 4);

Subject and substantive analysis of the studied object's state (its separate elements) and its semantically significant properties and features followed by the mapping in a toponymy of a symbolic representation or an attributive information on the electronic map;

Application of GIS Technologies in Historic ... 153

Fig. 3 Improvement of the Great Wall imagery's quality due to filtering out the spectral and atmospheric disturbances

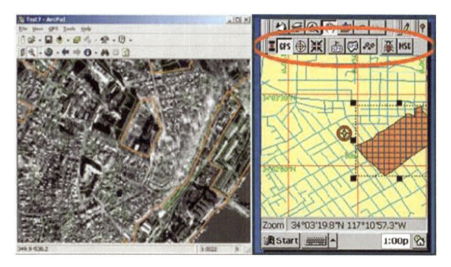

Fig. 4 Direct object's recognition and its recording on the electronic map

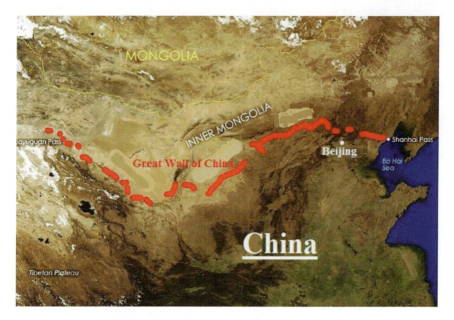

Fig. 5 Conditional and graphic mapping of the Great Chinese Wall on the generalized map in GIS

Conditional and graphic mapping of the being studied object (its elements) on the integrating electronic map pertaining to an appropriate geographic area and accounting for the used format for the above electronic map (SXF, VPF, S-57 and other.). For the historic object used here as an example, the generalized representation can look like the one in Fig. 5.

At the final stage of the described program and technological process an integration of the attributive and subject data is performed via composing a valid script and testing the joint use of the matching GIS- applications and ERS data base by means of playing the separate actions that regard display of the conditional and graphic and photographic observations (data), partial research problems solving.

Technology intended for modeling retrospective dynamics of various historic as well as historic and geographic processes in many respects is similar to the above technology. This technology assumes a possibility of a sequential accumulation of snapshots' "time samples" for the space extended historic object and data of its mapping, at that, proving for a further space and time modeling of such images "playing' in a time sequential mode. The above gives a possibility to observe the time dynamics in the object's and its environment states, to evaluate a general tendency as well as its volatility in the object's variation.

In whole an integration of the cartographic geo-information and ERS data opens up qualitatively new possibilities in historic research, new technologic stratum for the historic information.

3 Technology for Rendering (Refinement) Boundaries of the Areas Having Ethnographic, Political and Economic Value

Conventionally the historic works use geographic maps for visual representation of the physical landscape underlying the progress in historic, ethnographic, political, economic and other processes as well as for reflecting their subject's sense and geographic interpretation. This point idea could practically be explained by any map of the historic and ethnographic nature. Figure 6 gives an example of such a map.

It should be stated that, as a rule, the regions' (areas) boundaries interpreting certain historic and ethnographic, historic and economic as well as other data are quite conditional.

Advanced GIS technologies expand significantly the possibilities of geo-interpretation for historic and ethnographic as well as for other data and serve as the toolkit in the correct formation (refinement) of the mentioned regions' boundaries. In particular a technology of so called "temperature maps" has found its appropriate application at the correct formation (refinement) of the regions' and areas' boundaries. The given title has been assigned to the used technology as a result of its certain similarity with a methodology defining the atmospheric fronts' zones based on the measures temperatures' maps. The idea of this technology consists in:

Some parameter of historic, ethnographic or other value with a given areal spacing at the representation (measurement) is mapped as an intensity of a definite color.

Fig. 6 The USSR map depicting the country regions matching the national composition of the population

Say, density of the Russian population may serve as an example of such a parameter with regard to Fig. 6;

The being interpreted parameter's values are defined by mathematic extrapolation techniques for the subareas that do not possess any measured data;

The being interpreted parameter's threshold value is defined with an allowance for the confidence probability, and its excess allows to refer the given geographic point (areal spacing, area) to the being detected region. For instance, if the Russian population density across the area exceeds 50 people per km^2 the area could be referred to the region of the Russian population primary settlement;

Software performs a colligation of all areas (i.e. of initial areal spacing) referred to the being detected region in accordance with the threshold decision rule.

Thus, the regions' (areas') boundaries interpreting historic and ethnographic as well as historic and economic and other data are derived based upon a unified quasi-objective procedure and their confidence level can be evaluated by the confidence probability value taken on at forming the threshold decision rule.

It should be noted that the given technology can play a role of a powerful research instrument for the advanced historians and ethnographers. Analysis of the modern research literature of historic and ethnographic nature, e.g., [3, 4], shows that a thorough retrospective study of a territorial dynamics in a course of the nations (people) ontogenesis is one of the most productive method in research history and ethnography. Consequently, the described technology is a technology of a well-grounded informatization and automation of this technique, and implies its wide application to the historic and ethnographic research.

4 Technology for Identification and Refinement of the Facts in the Development of Historic and Geographic Processes Based upon Visualization of Their Textual Descriptions

Geographically strong identification of various historic objects' locations, locations of some acts commitment, etc. performed based on two main location's coordinates, i.e., the latitude and the longitude does not match at all the object conditioned character of the geographic information transfer in the historic sources. This tendency is observed in the textual descriptions of the early stages' events when the geographic coordinates' notion had not existed. Also the strong geographic identification of locations is not pertinent for somewhat later historic textual and descriptive sources. The latest is caused by the fact that in practice people not involved in precise measurements of certain locations on the Earth surface broadly use so call qualitative notions in order to identify an object's location quickly and approximately (e.g., "In the area of Borodino village", "At the Cape of Good Hope"...). Such a representation of geographic information that is very natural for a human being allows for the quickest possible analysis of the object situation while solving real tasks in everyday practice. Naturally, this approach is specific enough for the historic sources.

Obviously, the method's inaccuracy at the locations' identification for various historic events and acts induces a possibility of errors in the process of stating these or those facts, as well as of incorrect historic interpretations, of intended or unintended adulterations.

General transition from a conventional "paper representation" of geographic maps to geoinformation systems and a steady intellectualization of GIS applications allow for furnishing the research historians with the coordinate information aimed at an object's location identification, as well as at interpreting the historic facts' textual descriptions in object-active form (i.e., in conventional qualitative categories of the activity's subject area) as an information intended for the location's identification.

The idea of such an identification for a historic event or object location consists in establishing a conformity of the object and natural denominations within the territorial and geographic polygons under certain discrecity. Then the identification is reduced to a trivial GIS task of the described location geographic point (point neighborhood) ingress into a matching polygon or the imbedded polygons system. So, for a "rough" positioning these polygons will be described by the categories like: "The Southern part of the Crimea Peninsula", "The Eastern part of the Gulf of Finland", etc. For the Earth surface areas described almost in detail the following nominative categories will be applicable: "To the North of the Msta bend", "To the North of Kotlin Island", "Koporye Bay", etc. For the definite geographic locations that are identified uniquely in accordance with the historic and documentary sources the categories will be: "Top of Sapun Mountain", "Stones to the South of Mostchny Island', etc. Obviously, the sizes of polygons', that match the sea areas described almost in detail and the definite geographic locations, are being determined through an expertise based upon an integrated generalization of geographic-terminological documentary sources of the considered historic period (i.e., configuration and denomination of the proposed polygons might be cardinally different on the same map when intended for the interpretation of different historic periods). As an example of practical realization of the proposed gradation Fig. 7 shows the map layout for the Gulf of Finland Eastern part. This layout consists in the polygons of two dimensional classes: the larger ones (thick border lines) and embedded small polygons (thin red border lines).

So, the subsystem of the intelligent support embedded in GIS allows to interpret a phrase taken from the historic archive source and adapted to the modern terminological spoken language (e.g., borrowed from [6]): "Two rowing galleys of the Russian Fleet took an engagement with the Swedish barque to the North of Seskar Island, having had put out the island and got it lost in the morning fog they faced the Swedish barque, that had been sailing from the Trabezon raid to the South. During the naval clash the barque had been put on fire, got burning and thrown to the Demanstay shoal by the Eastern wind" as some geographic square with rather narrow geographic coordinates: " Latitude 60 10.2 S; Longitude 28 43.5 W".

The described software technology for identifying the object and activity location envisages in cases of specific activities a possibility to aggregate a nomenclature of polygons' types, to alter approach to an analysis of historic facts within various polygons.

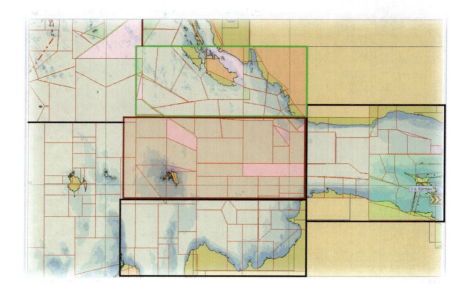

Fig. 7 Example of a territorial layout for the Gulf of Finland map (eastern part) aimed at location identification and facts of historic and geographic processes course

The above specification of location and geographic conditions in a course of these or those historic events allows to refine many "subtleties" in their descriptions, to state new or to refute doubtful facts. The best effectiveness such a refinement procures at a composition of the described technology and advantages of the modern geospatial-time systems for the processes course simulation.

Evidently, this technology does not bear an exhaustive capability for the technique being realized in historic studies and much depends upon the historic and geographic processes' description completeness, on the other hand it can be useful as an additional tool at the analysis of historic information under terms of its incompleteness, fuzziness and discrepancy.

5 Conclusions

Application of the intellectualized GIS in composition with advanced tools for information integration and fusion gives a possibility to enhance the scientific research efficiency in various spheres, including those related to humanitarian knowledge that definitely encompasses history, ethnography, anthropology and other.

While summarizing the features of new possibilities offered by an integration of methods and tools suggested by artificial intellectualization, information fusion and geoinformation technologies quite a number of the qualitatively new distinctions in intellectualized GIS oriented to historic studies.

The advanced intelligent geoinformation systems assume the following provisions:

- possibility to develop thematic maps of the subject area under study, that would be accompanied by the relevant visualization tools, specialized notations of the graphic symbols, matching editing aids, specialized ontologies and data types;
- visual development of the courses' models (courses' scenarios) for historic and geographic space processes aimed at performing GIS simulation;
- playing (simulation) of the courses' scenarios for historic and geographic space processes in real and arbitrary time scales accompanied by visual displaying of the symbols on the electronic map background;
- generation of instructions for the individuals making research decisions in cases when the course of scenarios' playing reveals obvious discords in the facts accountable at the research modeling, simulation games, and situations' analysis;
- user-friendly representation of geographic (space coordinated) data in the conventional qualitative categories of the activities' subject-area;
- intelligent analysis of the objects' activity in time and space, and other.

The proposed by the paper approach to the implementation of the intellectualized GIS technologies in historic and ethnographic research contemplates a possibility for various users' categories to arrange for parameterization and accretion of the courses' models (courses' scenarios) nomenclature in the area of historic and geographic space processes; for changing the discipline at analyzing positions on the electronic map, and, thus, allows to speak about the universality and broad scientific and research applicability.

References

1. Baranov YB (2009) Geoinformatics. Explanatory dictionary. GIS-Association Moscow, 204p
2. Popovich V (2003) concept of geoinformatic systems for information fusion. In: Proceedings international workshop information fusion and geographic information systems, St.Petersburg, Russia, pp 83–97, 17–20 Sept 2003
3. Loginov AV (2005) The power and the belief. State and religious institutions in the history and in the present. The Great Russian Encyclopedia, Moscow, 496 p
4. Loginov AV (2013) Russia and Eurasia. The Eurasian vector: searches for Russian civilization identity in the XX century. The Great Russian Encyclopedia, Moscow, 551 p
5. Gabriel J (2007) Situation management: basic concepts and approaches. In: Proceedings international workshop information fusion and geographic information systems, St.Petersburg, Russia, pp 18–33, 27–29 May 2007

6. Krotov PA (2011) Osudareva road of 1702 year Publisher. Historical illustration, St.Petersburg, 310 pp
7. Thill J-C (2011) Is spatial really that special? A tale of spaces. In: Proceedings international workshop information fusion and geographic information systems: towards the digital ocean, Brest, France, pp 3–12, 10–11 May 2011
8. Stottler H (2009) History and artificial intelligence. Oxford University Press. http://www.stottlerhenke.com/ai_general/history.htm

Geoinformation Systems for Maritime Radar Visibility Zone Modelling

Oksana Smirnova and Vasily Svetlichny

Abstract Today the problem of the radar visibility zone modelling under hydrometeorological conditions is essential. Using such modelling we can predict radar maximum capabilities for detection of various objects within the required environment in specific hydrometeorological conditions. Obtained information allows not only to provide the high efficiency of radar operation but to improve its work, save material and energy resources. With the active development of GIS and environmental monitoring systems (including space system) accounting more fully for hydrometeorological conditions such as atmospheric precipitates, heaving of the sea and other becomes realizable. However it is required to improving the mathematical methods of the operational creation of the radar visibility zone with hydrometeorological conditions.

Keywords Geo-modelling · Radar visibility zone · Attenuation function

1 Introduction

At present time different types of radars are used as means of detecting various objects at sea. They are designed for observation of dynamically changing above-water and air situation in given area, as well as for solving problems of short term and long term prediction of visibility zones with consideration of hydro-meteorological conditions. Main complications of operational evaluation of radiolocation visibility are associated mostly with calculation of radar attenuation function in

---

O. Smirnova (✉)
St. Petersburg Institute for Informatics and Automation of the Russian Academy of Sciences (SPIIRAS), 39, 14 Linia VO, St. Petersburg 199178, Russia
e-mail: sov@oogis.ru

V. Svetlichny
SPIIRAS Hi-Tech Research and Development Office Ltd., St. Petersburg, Russia
e-mail: svetlichniy.va@gmail.com

© Springer International Publishing Switzerland 2015
V. Popovich et al. (eds.), *Information Fusion and Geographic Information Systems (IF&GIS' 2015)*, Lecture Notes in Geoinformation and Cartography,
DOI 10.1007/978-3-319-16667-4_10

diffraction zone, adjacent to the sea surface. Methods of calculating radiolocation signal distribution developed as of today allow to solve partial radiolocation problems, such as calculation of single or group radar range, situated on ships or on shore, creation of radio range circle diagram, building of detection zones on cartographic under-layer and other [1].

Undoubtedly, most convenient platform for solving operational prediction problems of radiolocation visibility and visualisation of results is geoinformation systems (GIS). They provide effective processing of large sets of initial data that have geospatial association and also operative visual representation of obtained results on cartographic under-layer.

Thus, development of new mathematical methods of operational prediction of maritime objects' radiolocation visibility with consideration of hydrometeorological conditions in location region, radio signal propagation environment, noise, and above-water and air situation is needed.

These methods and their realisation in complex of radiolocation calculations will allow to considerably elevate the efficiency of solving target detection problems, their coordinates and motion parameters, air-to-air defence problems of the ship, using unmanned aerial vehicles, helicopters and airplanes of deck-based deployment etc.

2 Problem Statement

Different ships, airplanes, helicopters are the main objects under observation in the radar calculation system. Generally radars are combined into systems that allow to keep under control large areas. The upside of using the radar groups is the improvement of characteristics of the object observing system due to the reciprocal exchange of information between radars. Therefore the radar calculations system must allow to receive information both from single and group radars.

Important part of the radar system is the external environment subsystem. External environment of the radar functioning is above-water, under-water and air spaces that contain objects and where radio waves are propagated, providing the location of objects. Besides, external environment includes noise sources: passive noise (non-homogeneity propagation environment), radar clutter, and also objects, which negatively effect elements of the radar, not only of natural but also of synthetic origin.

External environment description includes:

- individual characteristics of objects' location and noise sources;
- description of the predicted above-water, under-water, air and noise conditions;
- description of the radio wave propagation environment for single or group radars.

Each radar is characterized by its operating conditions, such as radar observability, weather conditions, bottoming surface type, active noise. The following information is distinctive for objects under observation: radar cross-section, speed, statistical model for the return signal (Marcum model, Sverling model).

Under the radar clutter we imply the reflection from objects not targeted by radar. Radar clutter sources are earth surface, hydrometeors, state of the sea, synthetic metal clutter (from hull of the ship), dust clouds, big birds and others.

Effect of the wave propagation environment on radar manifests in the form of radio attenuation and bend of its propagation path (radio refraction).

All this information is closely related to spatial data, therefore it is expedient to apply GIS-technologies, in particular the principles of the geoinformation modelling, for efficiency and clearness of the visibility zone representation.

On the basis of the mathematical methods and associated calculations of the radar visibility characteristics we need to make single radar or group radar visibility zone modelling in given location region in accordance with hydro-meteorological and noise situation. By a radar visibility zone we here mean a part of the sea or air surface within which it is possible to observe location of objects with specified reflective properties and with specified probability of detection.

Also we have developed algorithms for solving this problem and applied them to the radar calculations system.

3 Generalized Method of the Radar Visibility Zone Building

According to the polar duality principle [2], in the generalized conic coordinate system, field of any given currents in some stratified medium can be described using two single-component vector Hertz's potentials of electric (TM) and magnetic (TE) type, sources received from total current in antenna using special procedure. Derivation of the integral equations for attenuation function of TM- and TE-fields are carried out for planar stratified troposphere in Descartes coordinate system considering earth curvature in parabolic approximation. TM- and TE- Hertz's potentials satisfy simple heterogeneous Helmholtz equations (complex values are marked by prime):

$$\Delta \Pi_{_s}^{'}(\vec{r},z)+k_0^2\varepsilon'(z)\Pi_{_s}^{'}(\vec{r},z)=-\frac{i}{\omega\varepsilon_0}\left(j_z^{'}-\frac{\partial Q'}{\partial z}\right);$$

$$\Delta \Pi_{m}^{'}(\vec{r},z)+k_0^2\varepsilon'(z)\Pi_{m}^{'}(\vec{r},z)=-M'(\vec{r},z). \qquad (1)$$

Here $\varepsilon'(z) = 1 + 2N(z) + 2\frac{z}{a}$ and $N(z)$—are relative dielectric permeability and vertical refraction index profile (a—Earth radius); ($Q'\left(\vec{r},z\right)$ and $M'\left(\vec{r},z\right)$—are potential functions of sources that are defined as solution of the equations:

$$\frac{\partial^2 Q'}{\partial x^2} + \frac{\partial^2 Q'}{\partial y^2} = \frac{\partial j'_x}{\partial x} + \frac{\partial j'_y}{\partial y}; \frac{\partial^2 M'}{\partial x^2} + \frac{\partial^2 M'}{\partial y^2} = \frac{\partial j'_x}{\partial y} - \frac{\partial j'_y}{\partial x} \qquad (2)$$

Electric and magnetic field vectors of TM- and TE- type are expressed as Hertz's potentials in the usual way [2].

Applying to Eqs. (1) and (2) Fourier transformation on horizontal coordinates x, y, excluding source functions and carrying items with $N(z)$ to the right part of equations, we reduce them to integral equations for Fourier image of potentials. Then, performing inverse Fourier transformation, we achieve integral equations for Hertz's potentials. Finally, we introduce attenuation functions for TM- and TE- field, $\Pi'_{e,m}(\vec{r},z) = \frac{\exp(ik_0 r)}{r} V_{e,m}(\tilde{x};\tilde{y},\tilde{y}_0)$, instead Hertz's potentials. And after some transformations we obtain single-type integral equations for attenuation functions:

$$V_{e,m}(\tilde{x};\tilde{y},\tilde{y}_0) = M_0 \int_0^{\tilde{x}} d\tilde{x}' \sqrt{\frac{\tilde{x}}{\tilde{x}'(\tilde{x}-\tilde{x}')}} \int_0^{\tilde{y} H} d\tilde{y}' V^{(0)}_{e,m}(\tilde{x}-\tilde{x}';\tilde{y},\tilde{y}') F_N(\tilde{y}') V_{e,m}(\tilde{x};\tilde{y}',\tilde{y}_0)$$
$$+ \frac{m}{4\pi k_0} \int_{\tilde{y}1a}^{\tilde{y}2a} d\tilde{y}' V^{(0)}_{e,m}(\tilde{x};\tilde{y},\tilde{y}') \int_{R^2} d^2 \vec{r}' \exp\left[-ik_0(\vec{r},\vec{r}')/r\right] \tilde{F}_{e,m}(\vec{r}',\tilde{y}'-\tilde{y}_0). \qquad (3)$$

Here

$M_0 = \sqrt{\frac{i}{\pi}} \left(\frac{k_0 r_3}{2}\right)^{2/3} N_0 \cdot 10^{-6}$—amplitude factor, $N_0 = N(0)$;

$\tilde{y} = k_0 z/m, \tilde{y}_0 = k_0 z_0/m$—scaled height of the observe point and antenna point above sea level;

$\tilde{x} = \frac{r}{a} m$—scaled horizontal distance to observe point;

$F_N(\tilde{y}) = N(z)/N_0$ function is expressed in terms of the variable $\tilde{y}$;

$\tilde{F}'_{e,m}(\tilde{x},\tilde{y})$—functions in the right part of Eq. (1) is expressed in term of the currents in antenna;

$V^{(0)}_{e,m}(\tilde{x};\tilde{y},\tilde{y}_0)$—functions in the right part of Eq. (1) is expressed in terms of the currents in antenna.

We assume that all troposphere irregularities, described by $N(z)$ function and effecting the radio waves propagation, are focused in height interval [0, H). In this height interval inequality (3) is integral equation, and for $\tilde{y} > \tilde{y}_H$ there is a computational formula allowing to calculate attenuation function if it has already been found by solving this equation in given area.

We need to know the distribution of antenna currents in order to gain extract formulas for attenuation field functions $V^{(0)}_{e,m}(\tilde{x};\tilde{y},\tilde{y}_0)$, but the problem of measuring them is beyond the scope of practical possibilities of a radar. It is possible to measure vector complex antenna directivity characteristics $\vec{F}(\vartheta,\varphi)$ in spherical coordinate system r, ϑ, φ centred on antenna and polar axis z, that is vertically up-directed. The influence of bottoming surface on the current distribution for highly raised radar antennas is negligible and we can use the relationship between Fourier images of sources in Eq. (1) on all three coordinates x, y, z and vector directional characteristics:

$$\widetilde{\vec{F}}_s(k_0\vec{r}/R, k_0z/R) = iconst \frac{(\cos\vartheta\vec{e}_r - \sin\vartheta\vec{e}_z, \vec{F}'(\vartheta,\varphi))}{\sin\vartheta};$$

$$\widetilde{\vec{F}}_m(k_0\vec{r}/R, k_0z/R) = iconst(\vec{e}_\varphi, \vec{F}'(\vartheta,\varphi)),$$
(4)

where $R = \sqrt{r^2 + z^2}$. The constant in (4) is defined by the radar power. As these formulas demonstrate, the vector complex directional characteristics do not allow to find sources in Eq. (1), since it defines Fourier images of these sources only on the surface of the sphere with radius k_0 in three-dimensional wave vector space. However, on rather long distance from radar (including semi-shadow and shadow zone) attenuation field functions $V_{e,m}^{(0)}(\tilde{x};\tilde{y},\tilde{y}_0)$ are linearly dependent on Fourier images of the source functions, the latter are taken for special angle value $\vartheta = \pi/2$. Thus, if we know the complex vector directional characteristics, we can find complex weight coefficients that determine contribution to TM- and TE-field in total field in these zones. Azimuthal parts of the antenna pattern in the attenuation functions $V_{e,m}^{(0)}(\tilde{x};\tilde{y},\tilde{y}_0)$, $V_{e,m}(\tilde{x};\tilde{y},\tilde{y}_0)$ for each (e, m) types act as weight coefficients and further will be omitted. As a result we turn to easier integral equations for attenuation functions:

$$V_{e,m}(\tilde{x};\tilde{y},\tilde{y}_0) = V_{e,m}^{(0)}(\tilde{x};\tilde{y},\tilde{y}_0) + M_0 \int_0^{\tilde{x}} d\tilde{x}' \sqrt{\frac{\tilde{x}}{\tilde{x}'(\tilde{x}-\tilde{x}')}} \int_0^{\tilde{y}H} d\tilde{y}' V_{e,m}^{(0)}(\tilde{x}-\tilde{x}';\tilde{y},\tilde{y}')$$
$$\times F_N(\tilde{y}') V_{e,m}(\tilde{x};\tilde{y}',\tilde{y}_0).$$
(5)

It should be noted, however, that relations (4) are valid in the far region of antenna and thus are approximate: they do not contain small diffraction refinements to account for which we need to know the currents in the radar antenna. Special calculations are made for the source in shape of horizontal electric dipole (currents there are known). Potential functions' apparatus $Q'(\vec{r},z)$, and $M'(\vec{r},z)$ leads to correct formulas for field in diffraction zone that for given source includes both TE- and TM-contribution [4].

If neither the weight coefficient of the attenuation function of TM- and TE-field in the homogeneous troposphere are included in integral equation (3) this functions have usual representation [4]:

$$V_{e,m}^{(0)}(\tilde{x};\tilde{y},\tilde{y}_0) = 2\sqrt{i\pi\tilde{x}} \sum_{n=1}^{\infty} \frac{\exp(i\tilde{x}t_n)}{t_n - q_{e,m}^2} \frac{w_1(t_n-\tilde{y})}{w_1(t_n)} \frac{w_1(t_n-\tilde{y}_0)}{w_1(t_n)}$$
(6)

In the illuminated zone in integral equation (5) we can go from reduced coordinates $\tilde{x}, \tilde{y}, \tilde{y}_0$ to usual x, z, z_0, Eq. (5) will appear as:

$$V_{e,m}(x; z, z_0) = V_{e,m}^{(0)}(x; z, z_0) + \sqrt{\frac{ik_0^3}{2\pi}} \int_0^x dx' \sqrt{\frac{x}{x'(x-x')}} \int_0^H dz' V_{e,m}^{(0)}(x-x'; z, z') N(z') V_{e,m}^{(0)}(x'; z', z_0) \quad (7)$$

Considering the typical heights of the radar antenna location and their operating wavelengths the attenuation function $V_{e,m}^{(0)}(x; z, z_0)$ in the illuminated zone (but in far region in relation to radar antenna) is represented by simple equation:

$$V_{e,m}^{(0)}(x; z, z_0) = D(x, z-z_0) \exp\left[\frac{ik_0(z-z_0)^2}{2x}\right] + \frac{(z+z_0) - \delta_M x}{(z+z_0) + \delta_M x} D_*(x, z+z_0) \exp\left[\frac{ik_0(z+z_0)^2}{2x}\right] \quad (8)$$

where $D(x, z-z_0)$ and $D_*(x, z+z_0)$—are coefficients of the radiation pattern antenna and its mirror view in the bottoming sea surface.

Using formula (8), we can get approximate analytical solution for integral equation (7) at the initial section of radio wave propagation path. For this we write solution of this equation in the form of Neumann series: $V_{e,m}(x; z, z_0) = \sum_{n=0}^{\infty} V_{e,m}^{(n)}(x; z, z_0)$ and take into account recurrence relation between successive terms of this series:

$$V_{e,m}^{(n)}(x; z, z_0) = \sqrt{\frac{ik_0^3}{2\pi}} \int_0^x dx' \sqrt{\frac{x}{x'(x-x')}} \int_0^H dz' V_{e,m}^{(0)}(x-x'; z, z') N(z') V_{e,m}^{(n-1)}(x'; z', z_0)$$

Integral with respect to z' we calculate successively in terms of the Neumann series with the first items calculated by method of stationary phase. Function $N(z)$ and preexponential factors in the formula (8) are considered slowly varying functions. Using majorant evaluation method we can prove that obtained Neumann series converge at all x, z. More than, in the consequence of some transformation we can find analytical formula for the found series sum:

$$V_{e,m}(x;z,z_0) = D(x,z-z_0)\exp\left\{\frac{ik_0(z-z_0)^2}{2x} + ik_0x\int_0^1 d\tau N[|(z-z_0)\tau+z_0|]\right\}$$
$$+ \frac{(z+z_0)-\delta_M x}{(z-z_0)+\delta_M x} D_*(x,z+z_0)\exp\left\{\frac{ik_0(z+z_0)^2}{2x} + ik_0x\int_0^1 d\tau N[|(z+z_0)\tau-z_0|]\right\}.$$
(9)

Obtained result has simple physical interpretation. The second items in exponents (9) are additives to phase of the direct and indirect rays from bottoming surface, radiated by antenna in observer point, caused by inhomogeneity of troposphere.

Integral equation with respect to x or attenuation function relates to Volterra equation class, therefore for finding solutions we use numerical next-step method on the basis of the quadrature formulas for integrals. These formulas take into account root particularity on both ends of interval of integration. Consequently, numerical next-step solution algorithm of integral equation (5) is described by the next formula:

$$V_{e,m}(\tilde{x};\tilde{y}_k,\tilde{y}_0) = \frac{\pi M_0 \sqrt{\tilde{x}}}{N_1}\sum_{n=1}^{N_1}\sum_{l=1}^{N_2} w_l V_{e,m}^{(0)}(\tilde{x}-\tilde{x}'_n;\tilde{y}_k,\tilde{y}'_l) F_N(\tilde{y}'_l) V(\tilde{x}'_n;\tilde{y}'_l,\tilde{y}_0)$$
$$+ V_{e,m}^0(\tilde{x};\tilde{y}_k,\tilde{y}_0)$$
(10)

where $\{\tilde{x}_n\}, \pi/N_1$—are nodes and weights of the quadrature formula for calculation of outer integral in (5), $\{\tilde{y}_k\}, \{w_k\}$—the same for inner integral, N_1, N_2—orders of the quadrature formulas. On the initial section of propagation path we apply function (9) as integral equation solution, and as $V_{e,m}^{(0)}(x;z,z_0)$—(8), which are replaced by function (6) in semi-shadow zone.

4 Method Realization Based on GIS

Based on introduced mathematical apparatus, complex of radar calculations with the use of GIS technologies was developed.

Suggested complex of radar calculations includes the following:

- GIS-interface;
- geo-information modelling subsystem;
- library of mathematical functions that includes means of mathematical modelling of radar visibility zones and calculating of radar observation characteristics;

Traditional part of GIS—is GIS interface, intended for visual representation of special data in different geographical numerical formats and objects, and adapted to depicture data from radar calculations.

Modelling subsystem is designed to fulfil geo-information modelling and prediction (short term and long term) of radar visibility zones in specific radar functioning conditions with consideration of hydrometeorological situation in location region.

It should be noted that the radar field, obtained based on formula (10), depends on location of radar source, since values of the attenuation function in each point of signal propagation are unique. The various phenomena, caused by influence of meteorological conditions, influence radar field configuration. Such phenomena are:

- radio-wave refraction in troposphere;
- radio-wave attenuation and scattering by hydrometeors. Effects of refraction are taken into account as:
- attenuation factor calculation while radio-wave passes through the troposphere;
- intensity of masking layer created by back-reflections from various hydrometeors (snow, rain, fog) contained in troposphere. Moreover, the known vertical profiles of index refraction along the propagation path of radio waves are taken into consideration.

Proximity and curvature effects of earth appear in the interference and diffraction of radio-waves and they are taken into account by means of radar attenuation function.

The modeling subsystem includes:

- module for calculating range of detection air and above-water objects for individual radar. Range of first sighting of the target is calculated with given probabilities of correct detection and false alarm, also the means and root- mean-square deviations of detecting air and above-water objects are obtained;
- module for calculating radio-locating stations' visibility zones, joined into the unified system, with information exchange on target level. Visibility zone can be built both for individual radar and for a group of separated radars. Limit of visibility zone is a range of detection air and above-water targets, calculated with given probability of correct detection with consideration of active and passive noise;
- module for calculating the density distribution of electromagnetic field power flow, created by emitting radar, with consideration of radio attenuation while passing the troposphere [3, 4]. For solving problems of this class, given distribution of vertical radio wave refraction index is taken into account. Also the proximity and sphericity of earth surface, leading to interferential and diffraction radio wave distribution, is taken into consideration.

To considerably ease the complexity of calculating process, we use mathematical functions library for development of individual radio-location problems. It includes special mathematical apparatus for solving problems of modelling radar visibility zones and calculating radar visibility characteristics. Also library of mathematical functions encloses a set of functions of particular form (e.g., Airy functions, Bessel functions etc.).

Thus, GIS may be represented as some platform for the radar calculation and solution of following tasks:

- input of the initial modeling data including geographical binding of the radar source to the given spatial region;
- providing access to geospatial databases of troposphere parameters that define the parameters of the radar field;
- support of control interface simulation;
- output of simulation results in a user friendly form.

Figure 1 demonstrates the example of the GIS-interface which provides performance of various radar calculations. Using GIS-interface, user can choose the observation region, observation objects and observation means for these objects. Observation objects and means can be mapped both manually by the user with associated tool bars or automatically using united object ontology.

Objects, necessary for radar calculations, are applied to radar plot according to their bearings and distances. Radar plot interface allows to correct location of the carried objects.

The form shown in Fig. 2 allows users to set necessary characteristics: the active and passive noise and their characteristics, the radar characteristics of the condition of the radio wave propagation over wave-covered water, radar specification (Fig. 3).

The results of tasks solution are the following:

- target range of detection with specified probability of correct detection and missed detection and also the mean range and root- mean-square deviation from range of detection (Fig. 4);

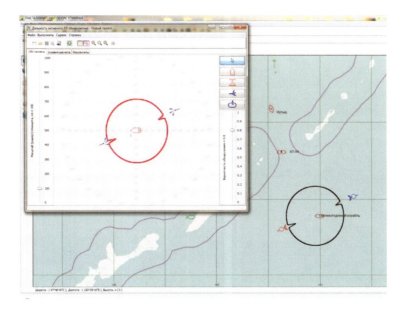

Fig. 1 GIS-interface of the radar calculation system

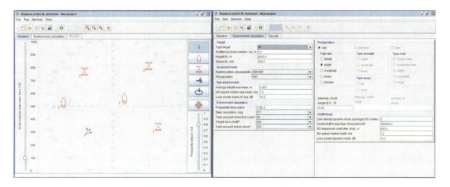

Fig. 2 Initial conditions for radar calculations

Fig. 3 Radar specification

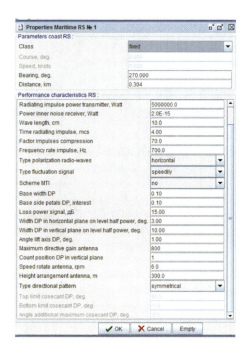

- detection zone for initial conditions for each variant. Zone borders are range of detection with installed probability of correct detection;
- graph of the probability of correct detection for several radar scans depending on range of detection according to selected bearing;
- distribution of the attenuation function in vertical plane and density power from emitting radar in illuminated and diffraction zones in term of color chart (Fig. 5). The user can obtain the cross- section of the radar field value in any point of presented zone.

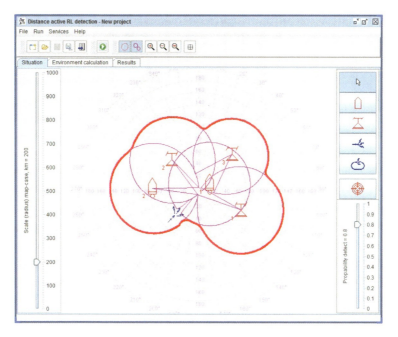

Fig. 4 Task solution results with according of the active and passive noise

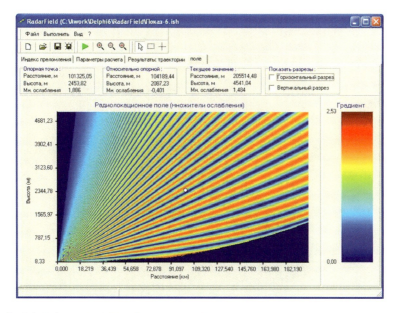

Fig. 5 Calculation radar field (radio attenuation factor)

5 Conclusion

The introduced method of calculating characteristics of radio signal distribution, based on integral equation for attenuation function for every given parameters of radio antenna, allows to solve problems operational predicting radio visibility. Derived mathematical results are used in the complex of radio-locating calculations and are displayed with the use of GIS-interface.

Complex of radio-locating calculations can be implemented in tactical situation control systems, stationary systems and for mobile carriers that include radars.

In the future, we intend to pay close attention to development of methods of calculating characteristics of radio signals distribution with consideration of hydrometeor influence based on intelligent geoinformation technologies.

References

1. Makarov GI, Novikov VV, Rybachek ST (1991) Electromagnetic waves propagation over earth surface. Nauka, Moscow, 196 p (in Russian)
2. Popovich VV, Leontiev YB, Ivakin YA, Smirnova OV (2013) The use of complex techniques radar calculations in radar systems. In: Proceedings of the scientific and technical conference «Radio-optical technology in instrument», Nebug, Russia, pp 64–81, 1–7 Sept 2013 (in Russian)
3. Smirnova OV (2014) GIS modeling of visibility zones of radar systems taking into account the meteorological conditions. In: Proceedings of the 7th Russian conference on control problem (ITU-2014), pp 302–306 (in Russian)
4. Fok VA (1970) Diffraction problem and electromagnetic waves propagation. Sovetskoe radio, Moscow, 517 p (in Russian)

Printed by Printforce, the Netherlands